About the Author

Dr. Manica Tomar is a dedicated plant pathologist with 15 years of experience safeguarding the health of fruits, forest plants, and flowers. A champion of integrated disease management, she has honed expertise in diverse areas, from field diagnostics to cutting-edge molecular techniques. Her commitment to advancing the field led her to the University of Toronto as a visiting fellow, where she delved into next-generation sequencing (including bulk RNA seq and data analytics) and state-of-the-art metabolomics. A prolific author, she has penned over 60 research papers published in journals of national and international repute, along with her first book (Progressive Plant Pathology]. With a passion for mentorship, she has guided nine postgraduate students and remains actively involved in research and teaching, currently focusing on disease management strategies for fruits and forest plants. Her practical experience and academic rigour makes her a valuable resource for anyone seeking to understand and combat plant diseases. This book is her brain child for the students seeking insights into the world of Plant Pathology.

Dr. Savita Jandaik a luminary in the realm of Mycology and Plant Pathology, shares expertise on edible and medicinal mushrooms, bridging the gap between science and sustainable utilization. She has made significant contributions to understanding the antimicrobial properties of higher fungi. Her pioneering research in fungal biotechnology has far-reaching implications for human health and environmental sustainability. Recipient of the prestigious Young Scientist and Women Scientist Fellowships from the Department of Science and Technology (DST), Government of India, Dr. Jandaik has made significant contributions to the field of plant pathology. Dr. Jandaik's groundbreaking work has been recognized nationally and internationally, and her research focuses on harnessing the potential of fungi for sustainable solutions in agriculture, healthcare, and environment. With a passion for unraveling the secrets of fungal-plant interactions, she has authored numerous research papers and chapters in esteemed journals and books. This book is a culmination of her extensive research experience and expertise in plant pathology.

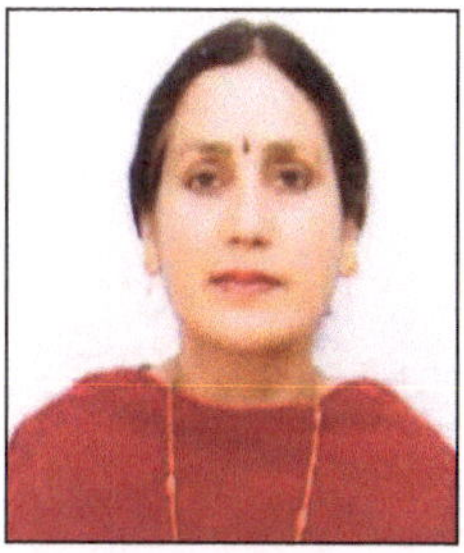

Dr. Sunita Chandel, a leading figure in plant pathology, has dedicated over three decades to safeguarding crops from disease. As and Ex Professor and Head of the Plant Pathology department at Dr. Yashwant Singh Parmar University of Horticulture and Forestry in Solan, Himachal Pradesh, she focuses her expertise on combating fungal diseases in fruit, vegetables and ornamental crops. Dr. Chandel's contributions to the field have garnered international acclaim, including prestigious fellowships from the Commonwealth Academic Staff and the Wageningen Institute in the Netherlands. Her groundbreaking research has been recognized with the Young Scientist Research Award from the DST in India and the CINADCO (MASHAV) Award from the Ministry of Agriculture & Rural Development in Israel.A prolific author and sought-after speaker, Dr. Chandel has published over 130 research articles and seven books, including the highly influential "Integrated Disease Management in Apple Orchards," a valuable resource for horticulturists across the region. Her leadership extends to over 15 research projects, securing funding from both national agencies like the Indian Council of Agricultural Research (ICAR) and international organizations like the Food and Agriculture Organization (FAO). Dr. Chandel's tireless efforts have not only advanced the science of plant pathology but also empowered countless farmers in Himachal Pradesh with the knowledge and tools to protect their livelihoods and contribute to a healthier, more sustainable agricultural sector.

Unlocking Plant Pathology

A Targeted Prep for ICAR-JRF and Agri Allied Services

Manica Tomar
Savita Jandaik
Sunita Chandel

NIPA® GENX ELECTRONIC RESOURCES & SOLUTIONS P. LTD.
New Delhi-110 034

NIPA® GENX ELECTRONIC RESOURCES & SOLUTIONS P. LTD.

101,103, Vikas Surya Plaza, CU Block
L.S.C. Market, Pitam Pura, New Delhi-110 034
Ph : +91-11-43860225, Mob.: +91 9717133558, 9540816132
E-mail: newindiapublishingagency@gmail.com
Website: www.nipaersources.com

Print ISBN: 978-93-58874-94-5

ebook ISBN: 978-93-58878-50-9

Composed and Designed by NIPA®.

Preface

The mind is not a vessel to be filled, but a fire to be kindled." - Plutarch

We are delighted to present "Unlocking Plant Pathology: Targeted Prep for ICAR JRF/SRF/ARS & Agri Allied Tests," a comprehensive resource designed to guide aspirants through the intricacies of plant pathology and help them succeed in competitive examinations. This book is the culmination of our teaching and research experiences, aimed at providing a structured and effective approach to mastering this crucial subject. Plant pathology is a dynamic and multifaceted discipline, requiring a deep understanding of fundamental concepts and their applications.

Recognizing the challenges faced by students preparing for competitive exams like ARS, Allied, JRF, and SRF, we have meticulously crafted this book to address their specific needs. This book offers a comprehensive blend of theory and practice. It includes concise yet detailed notes covering the core areas of plant pathology, ensuring a thorough understanding of the subject matter. We have included a substantial collection of question-and-answer sets to reinforce these concepts and assess comprehension. Furthermore, the book boasts an impressive bank of over 1000 MCQs, carefully curated to cover the breadth and depth of the syllabus. These MCQs are designed for rote learning and to stimulate critical thinking and problem-solving skills, essential for success in these competitive exams. We believe that this integrated approach of notes, Q&A, and MCQs will provide students with a robust foundation and the confidence to tackle any exam question.

This is the first part of our endeavor. A second part, covering the remaining topics within the plant pathology syllabus, is currently in the offing, and we anticipate its release soon. We are committed to providing a complete and up-to-date resource for all aspirants. We extend our sincere gratitude to the present and past doctoral students of the Department of Plant Pathology, College of Horticulture, Nauni and College of Horticulture and Forestry, Neri, for their invaluable assistance in formulating the MCQs. This book arose from the need to address students' unresolved queries and subject-related doubts.

While we have striven for accuracy and completeness, errors and omissions are always possible. We take full responsibility for any such shortcomings and welcome feedback from readers, which will be invaluable in improving future editions.

We hope that "Unlocking Plant Pathology" will serve as a trusted companion for students on their journey to achieving their academic and professional goals. We are confident that with dedicated effort and the right resources, success is within reach.

Authors

Contents

1

Brief Outline of Plant Pathology

Plant diseases have been documented since ancient times. Fossil data suggests that plants were impacted by illness 250 million years ago. The Indian Vedic text and Bible refer to diseases like rusts, mildews, and blights that have led to famine and significant economic impacts throughout history. Plant pathology or Phytopathology (Greek Phyto-plant pathos- suffering or ailment; logos- knowledge/study/discourse) is a division of agricultural sciences focused on the examination of plant diseases. It encompasses the recognition, comprehension, prevention, and management of plant diseases caused by different pathogens as fungi, bacteria, viruses, nematodes, and other microbes. **Plant pathology is the nature, cause and prevention of plant diseases.** **Plant disease** refers to the physiological functioning of the plants that are irritated by prolonged, phytopathogenic organisms biotic or infectious disease causing agents and abiotic or non-infectious causes such as edaphic, environmental, plant nutrition etc. covered under disorders. **Plant disease can also be defined as it is a malfunctioning process in plant body caused by continuous irritation induced by some causal agent that alters the physiology of the plants and results in some suffering or symptom production**. It may be other way stated that it is a sum total of the altered and induced biochemical reactions in plant system or part brought about by any biotic, mesobiotic agents or abiotic factor leading to malfunctioning of biological processes and ultimately manifesting at cellular and/or morphological level.

Plant Protection primarily focused on protecting crops from yield losses due to biological and non-biological causes. This problem is now aggravated and more challenging due to additional complexity generated resulting of limited natural resources that are available to agriculture that include water, arable soil, agricultural land, biodiversity, availability of non-renewable energy as well as human labour, fertilizers, pesticide etc.

Plant pathology is related to many other sciences viz., mycology, bacteriology, virology, microbiology, physiology, chemistry, biochemistry, genetics, biotechnology etc. that provide the knowledge required for the correct diagnosis and management of plant diseases.

The Importance of Plant Pathology can be Summarized as Follows

1 **Food security:** Plant diseases can cause significant crop losses, affecting food production and leading to food shortage. Plant pathologists play a crucial role in identifying and managing diseases to ensure sufficient food supply for the growing human population.

2. **Economic impact;** Plant diseases can cause substantial economic losses in agriculture. By studying the causes and methods of control, plant pathologists help farmers to mitigate losses and protect their livelihoods.
3. **Plant breeding and genetics**: Plant pathologists contribute to the development of disease-resistant crop varieties through breeding and genetics. By identifying resistance genes and understanding the interaction between plants and pathogens, they assist in the development of improved and sustainable crop varieties.
4. **Environmental sustainability:** Plant diseases can have negative ecological impacts, such as the spread of invasive species or the alteration of natural ecosystems. Plant pathology aids in managing and conserving ecosystems by minimizing disease outbreaks and their consequences.
5. **Sustainable pest management:** Plant pathologists study integrated pest management (IPM) strategies that minimize the use of pesticides and promote sustainable agricultural practices. They aim to control plant diseases while minimizing negative impacts on human health and the environment.
6. **Human health:** Some plant diseases can have direct impacts on human health. For example, mycotoxins produced by fungi on crops can contaminate food and pose risks to human health. By understanding and managing plant diseases, plant pathologists contribute in safeguarding the human health.

Out of the total average 36.5 per cent losses to the crops, it is estimated that 14.1 per cent is caused due to diseases, 10.2 per cent due to insects and 12.2 per cent by weeds. However, in recent years global crop yield loss estimated is 40% .Crop health is increasingly threatened by various factors like different pests, climate change and insufficient response of agricultural practices to these threats. It will have adverse impact on food chain, food security and human health and can put burden of crop loss across geographies and crops.

Overall, plant pathology is essential for the sustainable management of plant diseases, protecting crop yields, ensuring food security, preserving the environment, and promoting sustainable agricultural practices.

Examples of Plant Diseases Causing Huge Losses Worldwide

1. Cereal rusts
2. Cereal smuts
3. Ergot of wheat and rye
4. Brown spot of rice
5. Southern corn leaf blight
6. Powdery and downy mildew of grapes
7. Coffee rust
8. Dutch elm disease
9. Sigatoka leaf spot of banana
10. Sugarcane mosaic virus

11. Sugar beet yellows
12. Swollen shoot of cacao
13. Plum pox virus in stone fruits
14. Tomato spotted wilt virus
15. Citrus canker
16. Fire blight of pome fruits
17. Soft rot of vegetables
18. Peach yellows
19. Root knot nematodes
20. Sugar beet cyst nematodes
21. Dutch Elm disease
22. White pine blister rust

Objectives of Plant Pathology

Plant pathology has the following major objectives

1. To investigate the biotic, mesobiotic, and abiotic causes of diseases or disorders.
2. To investigate the processes of disease progression caused by infections.
3. To investigate the relationship between plants (host) and pathogens in relation to the environment
4. To create strategies for managing plant diseases.

How Disease Infections Shaped History of Mankind Some Classical examples

1. The Irish Potato Famine

Infection has significantly impacted human history over thousands of years. In the 1840s, the potato blight, demonstrated how an environmental disaster in a susceptible community can significantly impact human history. In the early 1840s, Ireland had the fastest-growing population in Europe before the potato blight struck. The Irish Potato Famine, also called *the Great Hunger*, started in 1845 due to a fungus called *Phytophthora infestans* (*P. infestans*) causing a devastating plant disease that quickly spread over Ireland. The infestation destroyed 50% of the potato harvest that year and approximately 75% of the crop during the following seven years. The potato blight had a devastating effect on Ireland and its inhabitants because the tenant farmers, who were under British administration, mainly depended on potatoes for food. The Potato Famine, which concluded in 1852, led to the deaths of approximately one million Irish individuals due to starvation and related factors. Additionally, at least another million were compelled to flee their country as refugees. This resulted in diaspora of about 80 million individuals to live in the continent of America. The diaspora had significant impacts on the social, economic, and political progress of the US.

2. The Stem Rust of Wheat

Cereal rusts have probably been a problem since the first cereal crops were grown in the Fertile Crescent. Spores of *Puccinia graminis* f.sp. *tritici* have been found in archaeological sites in Israel dating from 1300 B.C. Wheat, barley, and barberry all originated in the Fertile Crescent, so this complex relationship in the stem rust life cycle has an ancient history.

Figure 1: Courtsey APS

Wheat stem rust was a serious problem in ancient Greece and Rome. Rust was observed and recognized as early as the time of Aristotle (384-322 B.C.). The ancient Romans sacrificed red animals such as dogs, foxes, and cows to the rust god, *Robigo or Robigus*, each spring during the festival called the *Robigalia* in hopes that the wheat crop would be spared from the ravages of the rust (Figure 1). This festival was incorporated into the early Christian calendar as St. Mark's Day or Rogation on April 25. Historical weather records suggest that a series of rainy years, in which rust would have been more severe and wheat harvests reduced, may have contributed to the fall of the Roman Empire. European colonists introduced both wheat and barberry plants to North America. Barberry has been used for several practical purposes such as extracting a yellow dye from the bark, making jams and wines from the berries, crafting tool handles from the wood, and creating fast-growing, thorny hedges to contain animals. In Europe, farmers started to acknowledge the relationship between barberry plants and stem rust outbreaks in wheat. Barberry regulations were established in multiple New England colonies around the mid-1700s. Barberry continued to proliferate as early settlers expanded westward. After the devastating (1916 North American stem rust epidemic., a cooperative state and federal barberry eradication program was established in 1918. This program was partially motivated by the concern about food supplies during war.

3. Great Bengal Famine

The Bengal famine of 1943 was a significant catastrophe in twentieth-century South Asia. The event was catastrophic in terms of its magnitude, resulting in three million deaths. It took place during World War II, while India was under British colonial rule. The administrative inadequacies were immediately accountable for the destruction in Bengal. The primary cause of the famine was the brown spot disease of paddy caused by *Helminthosporium oryzae* Breda de Haan (*Cochliobolus miyabeanus*) Dreschler ex Dastur. No plant pathology literature has documented a disaster as severe as the Bengal famine of 1942-43. The only comparable event in terms of the impact on food crop loss owing to disease and the subsequent human tragedy is the Irish potato famine of 1845. As the crisis aggravated, social fabric disintegrated and families migrated to different cities. Thus Bengal famine can never be erased from the History of India and world as it is a true eyewitness to the wrath of plant diseases and indifference of the British Government that led to the death of millions by starvation and change in the course of mankind.

4. Coffee Rust

Coffee evolved as an under story plant in the mountainous woods of Ethiopia. Originally, it was utilized as a food source in compacted cakes, frequently transported by camel caravans operated by North African merchants. Initially used for medicinal and religious purposes, its invigorating and rejuvenating properties contributed to its widespread popularity as a beverage. By the early 1500s, coffee houses were prevalent in Egypt, Arabia, and Turkey, and European travellers began to enjoy this unique beverage. The Dutch recognized the commercial opportunities of coffee and started cultivating it in their territories in Ceylon, Sumatra, and Java. The term "Java" is still used as a colloquial term for coffee in several locations. A taste that began with the nobility and the wealthy soon drew in the common folk as well. Ceylon had been the leading coffee-growing region globally by the time the Dutch handed it over to the British in the 19th century. The British significantly increased the size of the plantations by clearing the island's forests to cultivate coffee on every possible land. In the 1870s, Ceylon's plantations were exporting up to 100 million pounds of coffee annually, with the majority going to England.

In 1867, coffee growers in Ceylon observed a "coffee leaf disease," which was later identified by Berkeley as being caused by a rust fungus. Berkeley named the species "*vastatrix*" to reflect the expected devastation from the early disease reports. Access a scanned copy of the page from the November 1869 Gardeners' Chronicle when Rev. M.J. Berleley first described *Hemileia vastatrix*. The path the fungus took from Ethiopia to Ceylon is still unknown. Initially, the coffee growers may have anticipated the issue to resolve swiftly, but by 1879, it became evident that the problem was persistent, causing desperation throughout the entire country. Harry Marshall Ward highlighted the dangers of extensive coffee cultivation without the protection of windbreaks to minimize the spread of rust spores. Ward suggested using a protective fungicide, lime-sulfur, to avoid infections. The coffee plantations' vigour and production decreased to the extent that they were no longer economically feasible. Ward arrived after the coffee was ruined, and his cautions about the risks of monocultures were ignored. After a time of significant economic and social turmoil, British planters transitioned from growing coffee to planting tea on a large scale, and British coffee consumers switched to drinking tea. Coffee rust quickly spread to India, Sumatra, and Java within a few years, causing the centre of coffee production to move to the Americas, which had not been affected by the rust at that time. Brazil quickly emerged as the primary coffee provider on a global scale. The Americas were able to keep coffee rust at bay until 1970, when it was found in Bahia, Brazil, due to a strict quarantine. The fungus spread rapidly through the coffee-growing regions of South America and Central America within ten years, as most of America's coffee originated from a single plant vulnerable to rust. *H. vastatrix* is present in almost all coffee-producing regions worldwide, excluding Hawaii.

5. Downy Mildew of Grapes

Grapevines are a prevalent fruit crop worldwide, with extensive cultivation in Europe, North and South America, South Africa, and Australasia. Grapes are utilized in the manufacturing of wine, brandy, or non-fermented beverages and are consumed fresh or dried as raisins. Resistant grape rootstocks from North America were brought to Europe in 1878 to manage the root-feeding aphid *Phylloxera*. Downy mildew, caused by *Plasmopara viticola*, is a major disease of grapevines. The downy mildew disease was likely introduced into France at the same time, most likely as oospores on imported rootstocks, and then spread extensively over Europe. P.M.A. Millardet originally utilized Bordeaux mixture in 1885 to manage downy mildew. A farmer reportedly used a mixture of copper sulphate and lime on grapes beside the roadside to create a chemical deposit that would deter theft by anyone passing by. Millardet's thorough experimentation resulted in the creation of Bordeaux mixture, the initial widely adopted chemical for safeguarding plants against fungal infections. The description of this work has been translated into English and is referred to as a Phytopathological Classic. A statue of Millardet was installed in a park in Bordeaux as a gesture of appreciation from the French populace for his significant discovery. Bordeaux combination continues to be a significant fungicide in modern times.

6. Ergotism

It is the effect of long-term ergot poisoning, traditionally due to the ingestion of the alkaloids produced by the *Claviceps purpurea* fungus that infects rye and other cereals, and more recently by the action of a number of ergoline-based drugs. It is also known as **ergotoxicosis, ergot poisoning** and **Saint Anthony's Fire**. In the Middle ages, the gangrenous poisoning was known as "holy fire" or "**Saint Anthony's fire**", named after monks of the Order of St. Anthony who were particularly successful at treating this ailment."Ergotism", in its modern sense, was first recorded in 1853; however, Sir Thomas Browne in his Christian Morals (published posthumously in 1716, although penned sometime in the 1670s) also made mention of 'ergotisms'. Notable epidemics of ergotism occurred up to the 19th century. Fewer outbreaks have occurred since then due to rye being carefully monitored in developed countries.

7. Other Important Epidemics

Panama disease of banana, caused by *Fusarium oxysporum* f.sp. *cubense* is one of the six most destructive diseases recorded in the history of plant diseases which destroyed around 4 million plants in Jamaica alone with 20 years of its appearance. Citrus tristeza or quick decline of citrus caused by *Citrus tristeza virus* destroyed millions of citrus trees in the world.

Important Definitions in Plant Pathology

Disease Definition of disease given by Horsfall and Diamond (1959) and accepted by British Mycological society and American Phytopathological Society is "Disease is a malfunctioning process caused due to continuous irritation which results in suffering producing symptoms in plants".

Disorder Plant diseases due to environmental conditions or other abiotic factors. eg. Khaira disease of rice due to Zn deficiency, black tip of mango due to Sulphur dioxide toxicity, whip tail of cauliflower due to Molybdenum deficiency.

Pathogen An entity that incites disease or any living agent that makes the plants to suffer (pathos) eg., protozoa, chromista, fungi, animalia, bacteria, phytoplasma, virus, viroid,virusoid.

Etiology The study of causes and origin of diseases is called etiology.

Parasite (Gr. *Parasitos*) An organism that lives in or on the host plant at certain stages of its life cycle to draw nutrition without affecting the essential functions of the host. eg., Cuscutta, Dodder, Rhizobium, lichens, mycorrhizae, Head lice, Malarial parasite, Bed bugs. Hence all pathogens are parasites but all parasites are not pathogens.

Sparophytes/saprobes Are the organisms which derive their nutrition from the dead organic matter. Some parasites and saprophytes may have the faculty (or ability) to change their mode of nutrition.eg Mushrooms, molds, *Rhizopus, Penicillium, Mucor,* and *yeast.*

Facultative saprophytes Are ordinarily parasites which can grow and reproduce on dead organic matter under certain circumstances. They are also called hemibiotrophs which attack the living tissues in such a way as biotrophs, but continue to grow and reproduce after the tissues is dead. eg *Botrytis cinerea, Rhizoctonia, Pythium* sp.

Biotrophs Are the organisms which regardless of the ease with which they can be cultivated on artificial media obtain their food from living tissues only in nature in which they complete their life cycle. They were earlier also called obligate parasites, e.g. rusts, smuts, powdery mildews, etc.

A parasite is called **necrotroph** when it kills the host tissue in advance of penetration and then lives saprophytically. The necrotrophs are also known as perthotrophs or perthophytes.e.g. *Sclerotium rolfsii* and Pythium species.

Similar to necrotrophs are **facultative parasites** which live as saprophytes but under favourable conditions they can attack living plants and become parasites eg *Armillaria, Ganoderma, Fomes* fungi causing root and butt rot in trees.

Pathogenicity Is the ability of a pathogen to cause disease under a given set of environmental conditions.

Pathogenesis Is the chain of events that leads to development of a disease in the host.

Parasitism Is a phenomenon by which a plant parasite becomes intimately associated with the plant; it draws nutrition and multiplies and grows at the expense of the plant host.

Virulence Is a measure or degree of pathogenicity of an isolate or race of the pathogen.

Aggressiveness Is often used to describe the capacity of a pathogen to invade and grow in the host plant and to reproduce on or in it. This term like virulence is used as measure of pathogenicity.

Immunity Immunity of a plant against a disease is absolute quality. It denotes the freedom of plant from disease, when the pathogen cannot establish parasitic relationship with the host. High resistance and low susceptibility approach immunity.

Disease resistance Is the ability of an organism to overcome completely or in some degree the effect of a pathogen or other damaging factor; whereas susceptibility the inability of the plant to resist the effect of the pathogen or other damaging factor.

Hypersensitivity Is the extreme degree of defense in which there is rapid death of the cells in the vicinity of the invading pathogen. It halts the further progress of the pathogen. Thus, hypersensitivity is a sign of very high resistance approaching immunity.

Infection Is the establishment of the parasitic relationship between the pathogen and host following entry or penetration.

Incubation period Is the time elapsing between penetration and completion of infection i.e. development of the disease symptoms.

Invasion and colonization Is the growth and multiplication of the pathogen through the tissue of the host to varying extent.

Pathovar Is a bacterial strain or set of strains with the same or similar characteristics, that is differentiated at infra- subspecific level from other strains of the same species or subspecies on the basis of distinctive pathogenicity to one or more plant hosts eg. *Psuedomonas syringae* pv. *syringae, P.s* pv. *avenae*

Race Race is used to refer to groups of host genotypes that allow for the characterization of virulence. This means that parasitic fungi are grouped into races based on how strongly they affect certain host plants eg., *Cochliobolus heterostrophus* race T is a pathogen that causes Southern Corn Leaf Blight

Strain A pathogen strain is a species or strain of a pathogen that is transmitted between individuals and exploits a specific host niche.eg Covid virus strain delta.

Genesis of Plant Pathology-- A Historical Perspective

Humans have known that plants can become diseased, with subsequent losses in food quantity or quality, for centuries. Early references to plant diseases can be found in the Old Testament of the bible, in the book of Kings, for example, and early Indian, Sumerian and Roman references to plant disease can be found that date back to 2000 BC. The earliest descriptions of plant diseases are primarily descriptive. Our concepts of disease and pathogens did not exist at that time, and thus the causes of disease might appear strange in our minds. Diseases are elevated when plants are cultivated, so the history of plant diseases begins with the invention of agriculture and, by extension, with the prehistory of humanity. Plant diseases, like those of other living things, are physiological or morphological changes that occur throughout their lives. From then until the present day, it evolved primarily from the observations of multiple writers, such as the Greek Theophrastus in antiquity, Ibn al-'Awwām, Surapal in Vrikshayurveda or "the Sevillian," in the Middle Ages, and Alonso de Herrera in the Modern Era.

Ancient history of Plant Pathology

- **Rigveda** (3700 BC. Oldest mention of plant diseases like blight, powdery mildew, rust and tumors on trees and worship of Sun God and Agni God to avoid those. Two facts came out from verses in the Vedas
  - Visible and invisible creatures entered the body and caused the disease
  - Sun, heat and fire killed these poisonous creatures
- **Surapal** (1000AD. In his book "*Vrikshayurveda*" divided plant diseases in two categories Internal and External diseases. Internal diseases were caused by inroads of foreign organisms into the plant body. External diseases were attributed to non-parasitic injuries by heat, frost, wind, soil acidity, water stress and poor quality seed. For cure of both types of diseases Surapal prescribed the treatment of pits for planting trees and treatment of seeds etc.

- **Homer** (1000 BC. Mentioned the therapeutic properties of sulfur on plant diseases.
- **Albertus Magnus** (1200 AD. Conceived the idea that higher plant mistletoe was a parasite of plants which takes its food from the host plant and makes it sick.
- **Democritus** (470 BC. Recommended control of plant blights by sprinkling plants with olive grounds left after extraction of olive oil.
- **Sushruta** (400 BC. Made a mention about the plant diseases and suggested that "just as the proper season, good soil, water and vigorous seed produce a healthy plant".
- **Parashar** (400BC. Made a mention of powdery mildew and rust in his book *"Krishi Prasar"* and invoked the Wind God to move them away from his field.
- **Romans** (750 BC. Romans created two Gods Robigo and Robigus to save them from disease of wheat rust. Romans used to celebrate a festival called "Robigalia" on 25th April in which they used to sacrifice red dogs, foxes, and cows in attempt to please and pacify the so called God of Rusts *"Robigus"* so that it did not send rusts to destroy their crops.
- **Theophrastus** (384-322 BC. Wrote two books *"Historia Plantarum"* and *"De Causis Plantarum"* and mentioned about plant diseases like rusts, mildews and blights etc.) and expressed the opinion that these diseases were caused due to bad nutrition and bad air which occurred due to wrath of God.

Starting Dates of Fungi

1. **1729** - Italian botanist P. A. Micheli proposed fungi comes from spores; **Father of Mycology.**
2. **1755** - French botanist Tillet published a paper on bunt or stinking smut of wheat; discovered bunt is a disease of wheat.
3. **1807** - French scientist I. B. Prevost showed bunt of wheat is a fungus and showed evidence that a disease is caused by a microorganism.
4. **1821** - E. M. Fries published *Systema Mycologicum* for naming of fungi; he was named as **Linnaeus of Mycology.**
5. **1821** - Robertson of England stated that sulphur is effective against peach mildew.
6. **1845** - Irish Potato famine due to *Phytophthora infestans* caused starvation of million of people.
7. **1858** - J. G. Kuhn published first textbook in Plant Pathology – The Diseases of Cultivated Crops, their Causes and their Control.
8. **1861** -Anton de Bary (Germany) worked out the life cycle of potato late blight and first to prove experimentally *Phytophthora infestans* is the cause of potato late blight. He proved that fungi are causes but not the results of diseases. **He is the Father of Modern Plant Pathology.** He studied 'Heteroecism' in rusts *Puccinia graminis f.*sp. *triticina.* He also worked on involvement of toxins and enzymes in plant diseases.

His famous 5 students were Brefeld (German), Woronin (Russia., Millardet (French), Farlow (USA., Marshal & Ward (Britain).

9. **1918** -E.J.Butler published book on Fungi and Disease in Plants; he made exhaustive study on Indian fungi and the diseases caused by them. He is called as the Father of Modern Plant Pathology in India; He joined as the first Director of Imperial Bureau of Mycology (Commonwealth Mycological Institute, CMI), now CAB International Mycological Institute in Kew, England in 1920. He began the journal *Review of Applied Mycology*. He wrote 'Plant Pathology' in 1949 with S.G. Jones.

Phytopathological Classics These are the 15 classical and pioneering works of various scientists in the field of Phytopathology which laid the foundation of modern day science of Plant Pathology.Given below are the name of the scientists along with the name of their published paper and in the parenthesis are the names of the scientists who translated their work later in English.

1. **Fabricius** – 1774-Attempt at a Dissertation on the Disease of plants. (Raven)
2. **Fontana** – Observation on Rust of Grain 1767. (Pirone 1932)
3. **Millardet** – The discovery of Bordeaux Mixture 1885. (Schneiderhan 1933)
4. **Woronin** – *Plasmodiophora brassicae,* the Cause of the Cabbage Hernia 1878. (Charles Chupp 1934)
5. **Tillet** – Dissertation on the Cause of the Corruption and Smutting of Kernels of Wheat in the Head and on the me of preventing these untoward circumstances in 1755. (Humphrey 1937)
6. **Prevost** – Memoir on the Immediate Cause of bunt or smut of Wheat, and on the Prevention of Bunt 1807. (Keitt 1939)
7. **Mayer** (1886), **Ivanowski** (1892), **Beijerinck** (1898) and **Baurb** (1904) – Wrote three early papers on tobacco mosaic and one on infectious variegation. (James Johnson 1942)
8. **Berkeley** – Observations, Botanical and Physiological, on the Potato Murrain in1846, also includes selections from Berkeley's "Vegetable Pathology" made by the Plant Pathology Committee of British Mycological Society.1948
9. **Targioni Tozzeti** – True Nature and sad effects of the rust and other maladies of Wheat and of Oats in Field in 1767. (Tehon 1952)
10. **Bassi** – *Del Mal del Segno meaning "Of the diseases, of the Sign"* in 1835. (Yarrow1958)
11. **De Bary** – Investigations of the Fungi and the Diseases of Plants Caused by Them With Reference to Grain and Other Useful Plants 1853.(Arny and Moore 1969)
12. **Hartig** – Important Diseases of Forest Trees 1874. Gave a detailed report which established the microbial basis for wood decay. (Merrill et.al.,1975)
13. **Fischer/Smith** – The Fischer Smith Controversy "Are Their Bacterial Diseases of Plants?" Seven articles written in 1899 depict the classic and bitter debate regarding the existence of bacterial diseases of plants. (Cambell1981)

14.Ando/ Fukushi/ Storey – Viruses In Vectors: Trans-ovarial Passage and Retention. Classic papers on plant viruses and their insect vectors written in 1986.

15.Dutch Elm Disease – Original 11 articles that laid the foundation for studying the devastating disease of Elm trees in Europe. A Research led by 7 female Dutch scientists from 1920 – 1935.

Ancient Period

In India, the information on plant diseases is available in ancient literature such as rigveda, atharveda (1500-500BC., arthasashtra of Kautilya (321-186 BC., Sushrutasanhita (200-500AD., Vishnupuran (500AD., Agnipuran (500- 700AD., Vishnu dharmottar (500-700AD., etc. In Rigveda, not only the classification of plant diseases has been given, but the germ theory of disease was also advocated.

Mycological Plant Pathology

Pre-Modern Period (1600-1853)

Z Jensen 1590-1608

- Invented first compound microscope.

R Hooke (1635-1703)

- Microscope was used to prepare drawings of fungi. Made first illustrations of micro-fungi. Book "***Micrographia***" in 1655. Showed sporangia of *Mucor* and teliospores of rose rust (*Phragmidium mucoronatum*). Believed the spores were seed pods rather than seeds and fungi initially arose spontaneously.

Antony Van Leewenhoek (1673)

- Yeasts were observed with microscope.

R Morrison (1699)

- Fungi as a component of lichen were recorded by him.

E. Clusius(1601)

- He wrote "***Rariorium Plantarum Historia***", Described many mushrooms with the help of water colour drawings.

Van Sterbeeck (1675)

- Wrote "***Theatrum Fungorum of het Tooneel der Campernoelien***" in Dutch meaning "What are these fungi and where did they come from?" First book devoted entirely to fungi.

Buller

- He studied mechanism of spore discharge in Fungi (Buller phenomenon)

Carl Linnaeus (1753)

- Given binomial system for naming the plants & fungi.
- He published "*Species Plantarum*" . He is considered as "**Father of Botany**"

P.A. Micheli (Pier Antonio Micheli) (1729)

- He was first to know about the fungi and saw the spores of fungi.
- First person to grow uncontaminated culture of fungi growing artificially on.

- Melon pieces
- He demarcated the 1900 plants among a which 900 were fungi .
- He published a Monograph – "*Nova Plantarum Genera*" .
- He is "**Father of Mycology**"

Tillet (1755)

- Published a paper on bunt or stinking smut of wheat.
- Demonstrated experimentally that wheat seeds contained black powder on their surface produced more diseased plants than clean seeds.
- He emphasized that bunt was an infectious disease was caused by some toxin produced by the black powder.
- He reported that the chemical treatment of seeds with copper sulphate inhibited the contagious activity.

Fontana(1767) reported that fungi cause plant diseases.

Persoon, C.H. (1801)

- Given systematic of fungi and laid the foundation for taxonomy of fungi
- His Book – "*Synopsis Methodica Fungorum*".
- By binomial nomenclature classified various fungi as Uredinales, Ustilaginales and Gasteromycetes.

Prevost (1807)

- He proved that diseases are caused by microorganisms.
- First man to demonstrate microorganism as disease causing agent.
- He discovered the life cycle of "Bunt fungus".
- He observed the spore germination in bunt fungus.
- First man to suggest or use of copper sulphate for seed treatment to prevent seed borne disease.

Nees (1817)

- Introduced the term 'ascus' and afterwar dscoined the term 'spore'

Fries Elias Magnus (E.M. Fries) (1821-1832)

- He published "*Systema Mycologicum*".
- He regarded as "*Linnaeus" of Mycology* and "*Father of Systematic mycology*"
- He described rust and smut fungi as a product of the diseased plants.

Schwann (1837) recognized the fungal nature of yeast and named it "*Zuckerpilz*".

DGruby (1844) proved that fungi cause human diseases.

Tulasne brothers (L. R. Tulasne and C. Tulasne)

- ***'Selecta Fungorum Carpologia'*** in Latin(1861-1865)- three volumes
- Gave the concept of Polymorphism in fungi
- Showed relationship between conidial stages and their perfect fruit bodies
- L.R. Tulasne-**"Reconstructor of Mycology"**

Raulin (1867)

- Gave the concept of Mineral nutrition of fungi

Fuckel (1869)

- Created the group '*Fungi Imperfecti*'

Eriksson (1894)

- Demonstration of biological races in cereal rust

Modern Plant Pathology

Anton De Bary (1831-1888)

- Proved that the fungus *Phytophthora infestans* was the reason of late blight.
- He considered late blight of potato fungus, its nomenclature and name again it as *Phytophthora infestans*.
- He described the other diseases also such as rust, smut, downy mildews.
- Study the functioning of plant disease and reported the role of enzymes and
- poison in tissue degradation caused by *Sclerotinia sclerotiorum* in carrot.
- Described heteroecious behaviour of Rust fungi.
- Presented the terms Biotrophs, Facultative parasite and Facultative saprophytes.
- Reported the plasmodial stage of Myxomycetes

Authored two books

- Morphology and Physiology of Fungi, lichens and Myxomycetes (1867)
- Comparative Morphology and biology of the Fungi, Mycetozoa and Bacteria (1884)

For all his contributions regarded as

- Founder of "**Modern Experimental Plant Pathology".**
- **Father of "Plant Pathology"**

M. J. Berkeley (1855)

- Detected root knot nematodes in galls of Cucumis.
- Founder of Mycology in British - described many species of fungi.
- Described the Late blight of *Solanum tuberosum* pathogen, Rust of coffee and Grape downy mildew.

Authored books

- Outlines of British Fungology
- Introduction to cryptogamic botany

Julius Gotthelf Kuhn (1858)

He published a book in Plant Pathology entitled "The Diseases of Cultivated Crops, Their Causes and Their Control".

- Demonstrated the penetration of Bunt fungus into *Triticum* and its development in the plant
- Seed treatment by copper sulphate was given by him and brought into usage at farmers level.
- He described the Rusts, Smuts of cereals, ergot sooty mold and diseases of mustard.

Robert Harting (1839-1901)

Worked on Forest Pathology

He wrote two important books in German on

- Important Diseases of Forest Trees (1874)
- Text Book of Tree Diseases (1882)

He is called the "**Father of Forest Pathology**".

Brefeld & Its Colleague (1875-1912)

Discovered the systems of artificial culture of micro-organisms and gave "Pure culture Techniques"

Given the Single spore isolation technique

Described the life cycle of Cereal smut

Pierre Marie Alexis Millardet (P.M.A. Millardet) (1882)

He discovered the Bordeaux mixture for the prevention of downy mildew of *Vitis vinifera* caused by *Plasmopara viticola*

Described the root aphid *Phylloxera* of the *Vitis vinifera* vine and hybridization experiments in order to have resistant plants

M.S.Woronin (Russia)

Discovered the cause of Club root of *Brassica oleracea* (*Plasmodiophora brassicae*) Studied its life cycle

Proposed the elimination of diseased plants and burning them to destroy the resting spore

R. H. Biffen (1905)

He showed that resistance to yellow rust in wheat was hereditary and Follow Mendelian Method.

First to develop Rust resistance variety "Ribet" against yellow rust of *Triticum (Puccinia striiformis)*

Flor (1955)

- He gave "Gene for Gene Hypothesis" idea in Flax against Rust (*Melampsora lini*).

"For each gene to confer resistance in the host there is a equivalent gene in the pathogen that confers virulence and vice-versa".

Vander Plank (1963)

- He gave the idea of Horizontal and vertical resistance, Threshold theorem, Vertifolia effect, AUDPC (Area under Disease Progress Curve). He has written books on
  (i) Disease Resistance in Plants (1968)
  (ii) Principles of Plant Infections (1975).
  (iii)Plant Diseases Epidemics and Control (1963)
- Regarded as **Father of Epidemiology**

H.M.Ward (1901)

- Doing work on coffee rust and life description of pathogen.
- Explained the role of environment in coffee rust epidemics.
- Established the Hypersensitive method in Borme grass infected with brown rust.
- Wrote book Diseases in plants
- Father of **Tropical Plant Pathology**

J. H. Craigie (1927)

- First Discovered the function of Pycniospore of *Puccinia graminis tritici*
- Described the sexual stage of Rust fungi.

C L Shear & B O Dodge (1927)

- discovered red bread mould (*Neurospora crassa*. as an ideal organism to study genetics and law of heredity

Waksman (1942)

- He was the first to use the term 'Antibiotic'

E. C. Stakman

- Presented presence of physiological races in *Triticum* rust.
- Coined term Hypersensitivity.
- Wrote book Principles of plant pathology.

Blackslee

- Gave concept of Heterothallism in Fungi.

G Pontecorvo and J A Roper

- Observed parasexuality in *Aspergillus nidulans.*

J.C. Walker

- Described the role of Phenolic compound in plant defence machinery
- Described the resistance of coloured onion bulbs to *Colletotrichum circinans* was due to catechol and Proto-catechuic acid current in the dry scale
- He wrote two books on Plant Pathology and Diseases of Vegetable Crops .

Muller and Cruckshank

- He showed the involvement of Phytoallexin in plant disease resistance.

Katan et al. (1976)

- Gave first report on the use of solar energy for control of soil-borne diseases.

GC Ainsworth

- He is credited with the publication of the ***Dictionary of the Fungi*** along with G.R. Bisby in 1943. Wrote the book "***The Fungi An advanced Treatise***(Co edited with A Sussman and FK Sparrow).

Bacteriology

Anton Van Leeuwenhoek

- He was a "cloth merchant" in Delft Holland.
- He developed advanced magnifying lenses (300- 500 times magnification)
- He developed simple microscope
- Discovered the micro-organism like – Bacteria, Protozoa, algae and named them as "Little animalcules".
- He is known as **Father of 'bacteriological world**'.

Louis Pasteur

Proved the germ theory

Coined the terms Aerobic, anaerobic organisms

Proved fermentation is biological process

Developed Pasteurization process

Developed vaccine for Rabies disease

Regarded as "Father of Microbiology"

Robert Koch (1876)

- He showed that the anthrax disease of cattle is caused by *Bacillus anthracis.*
- He contributed Koch's rules or Koch's postulates of pathogenicity.
- Use of Gelatin as a solidifying agent
- Found causal organism of Tuberculosis disease.
- Known as "**Father of Medical bacteriology**".
- Regarded as **Father of bacteriological techniques**

Koch's Postulates

1. He postulated that micro-organism must be present in all diseased host.
2. The supposed causal agent must be isolated from the diseased host and grown in untainted culture.
3. The exact disease must be produced when a pure culture of the suspected causal agent is inoculated into a healthy vulnerable host.
4. The similar causal agent must be isolated again from the inoculated and infected host.

T. J. Burrill (1878)

- He first time proved that the fire blight of *Malus* & *Prunus* was caused by a *Erwinia amylovora.*

- First to Demonstrate Bacteria can cause plant disease

E. F. Smith (1900)

- Final proof for bacteria can infected plant
- Considered bacterial wilt of *Cucumis sativus*, Crown gall of *Malus*
- Provided Fourth Step of Koch's postulates
- Wrote books 'Bacterial diseases of plants', 'Bacteria in relation to plant diseases'
- Considered as **Father of "Phytobacteriology"**
- He first established that budding or grafting could be another method of spread of plant viruses.

Walther and Frau Hessey (1881)

- Used Agar –Agar as a solidifying agent in media.

R. J. Petri

- Established the Petri plates (two circular plates holding the media.

Doi and Ishi (1967)

- Discovered (MLO's) causing yellows type diseases (Mulberry dwarf, Aster yellows) constantly present in phloem of the plants.
- Ishi *et al.*stated that Tetracycline temporarirly disappeared the symptoms of MLO in plants

Davis and Worley, J.E. (1972)

- Discovered "*Spiroplasma*" in corn stunt disease.

Windsor, I.M. and Black, L.M. (1972)

- He detected first existence of Rickettsia like organisms in clover club leaf disease.

Virology

Carolus Clusius

- Recorded breaking of tulips due to viral infection

Adolf Mayer (1886)

- Discovered the tobacco mosaic virus disease.
- Named the symptom of this diseases as "Mosaic"
- Showed that the sap transmission of disease.
- He established that this disease (TMV) could be transmitted from diseased to healthy plants by the sap from leaves showing mosaic symptoms.

Dimitrii Ivanowski (1892)

- He demonstrated that tobacco mosaic virus (TMV) could pass through even those filters that retained bacterial cells. So he suggested that viruses were smaller than smallest known bacterium.

M.W. Beijerinck (1898)

- He determined that Tobacco mosaic was caused by "*Contagium Vivum Fluidum*" (infectious living fluid) which he called a "VIRUS".
- He discovered the causal agent of TMV.
- Coined the term 'Virus'
- Father of Plant Virology

Stanley, W.M. (1935)

- He purified and crystallized TMV.
- Reported Viruses are proteinaceous molecules
- He got Noble Prize.

F. C. Bawden and N. W. Pirree (1938)

- Proved that Viruses were Nucleoproteins.
- Showed that TMV consist of 95% protein and 5% Nucleic acid

Gierrer and Schramm

- Presented RNA of (TMV) is infectious non protein part.

Knoll and Ruska (1932)

- Invented electron microscope

Diener, T.O. (1971)

- Discovered the 'VIROIDS'
- Determined that potato spindle tuber disease is caused by small molecules of infectious ribonucleic acid (RNA. which he called "VIROIDS".

Stanley Prsuiner

- Discovered the Prions causing Scrapie disease in Sheep

A.N.Fire and C.C. Mello (1998)

- Work on RNA Interference gene silencing by ds RNA. Awarded Nobel Prize for this work in 2006.

Nematology

Nathan Augustus Cobb N.A. Cobb. (1913-1932)

- He contributed to describe nematode morphology, taxonomy and methodology.
- Given the term "Nematology".
- Specified techniques for separation of nematodes from soil - Cobbe Sieve technique.
- He is considered as "**Father of Plant Nematology**".

T. Needham (1743)

- He described first plant parasitic nematodes in wheat-galls-Ear cockle of *Anguina tritici.*

History of Plant Pathology in India

Surapal (11th century)

- Wrote *Vraksha Ayurveda*, first book in India having detail account on plant diseases & their control.

D. D. Cunnighham (1850-1875)

- Collection and Identification of Fungi of India
- Recognized the causal agent of Red rust of *Cephaleurous virescens* in Assam
- Studied the Rust and Smuts

K. R. Kirthikar

- First Indian Scientist to collect fungi and recognize in India

E. J. Butler (1905)

- First cryptogamist in India
- Worked on red rot of Sugarcane, Wilt of Red gram, Rice diseases
- Described the life cycle of smut diseases.
- Published book on "Fungi and Disease in Plants" in 1918.
- Monograph "*Pythiaceous and Allied Fungi*"
- Established Herbarium At IARI
- **"Father of Indian Plant Pathology"**.
- First Director of Imperial Mycological Institute in 1929

J. F. Dastur (1886-1971)

- First Indian plant Pathologist, to study in detail about Fungi and Plant diseases.
- He studied the genus *Phytophthora* and diseases of *Ricinus communis.*
- He is universally known for establishing the species of *Phytophthora parasitica* from *Ricinus communis.*

K.C. Mehta (1929)

- Worked on Epidemiology of Wheat rust
- He also demonstrated that *Barberry* has no role in spread of rust fungus in India.
- Discoved the cereal rusts in India.
- He wrote a monograph on Cereal Rust in India part-1 (in 1940) and part-2 in 1952.
- Regarded as Father of cereal Rusts in India.

B. B. Mundkur (1948)

- He worked on Management of *Gossypium* wilt through Resistance breeding
- Worked on Smut fungi, wilt diseases.
- He published a book "Fungi and Plant Diseases",
- Wrote Monograph Ustilaginales in India
- Started "Indian Phytopathological Society" in 1947 with its journal "Indian Phytopathology" in 1948.

M. J. Thirumalachar

- He introduced the usage of antibiotics in control of plant diseases e.g., *Aureofungin* and *Streptocycline.*
- Extensively studied the Smuts and Rust
- Wrote Monograph Uredinales of world, Cercosporae

J.C. Luthra and his Associates

- He recognized the solar treatments of *Triticum* seed for control of loose smut (seed borne disease).

T. S. Sadasivan

- Worked on biochemistry of Host- Parasite Relation
- Developed the concept of Vivo toxin
- Worked on the mechanism of *Gossypium* wilt (due to Fusaric acid.

S. N. Dasgupta

- Worked on biochemistry of Host- Parasite Relation
- Studied the Black tip of Mango

M. K. Patel (1948-51)

- Discovered the bacterial plant diseases in cotton and paddy.
- Proposed new family Phytobacteriaceae to include all pathogenic bacteria.
- Developed a synthetic medium known as Patel's Medium to culture of bacterium.
- Discovered many *Xanthomonas* and *Psuedomonas* species which are responsible for plant diseases.
- **Father of Phytobacteriology in India.**

Nene Y. L.

- Worked on pigeon pea sterility mosaic disease
- Described Zn deficiency of Khaira Disease of Paddy
- Published books on
    1. Fungicides in plant disease control
    2. Viral diseases of pulse crops in Uttar Pradesh.

G. S. Kulkarni

- A scholar of Butler, generated detail information on downy mildew and smut of jowar and bajra.

S. L. Ajrekar

- A student of Butler, considered wilt disease of *Gossypium*, sugarcane smut and ergot of jowar

Manoranjan Mitra

- He was considered as one of the greatest critical plant pathologists who worked on *Helminthosporium* sp. He first reported Karnal bunt of *Triticum* in 1931 from Karnal in Haryana.

T. S. Ramakrishnan

- A mycologist in Madras Government who cultivated diseased rye (ergot) for toxin production.
- He published two books entitled Diseases of Millets (1963) and Diseases of Rice (1971). Renowned plant pathologists viz., G Rangaswami and R. Ramakrishnan were his students.

2

Classification of Bacteria and General Characteristics

"Bacteria" = Plural form, "Bacterium" = Singular form

Bacterium is a single-celled microorganism that do not have cell organelles (mitochondria, endoplasmic reticulum, chloroplast etc.) or a true nucleus.Bacteria are microscopic, unicellular organism that belongs to the kingdom prokaryotes.

Morphology of Bacteria

Bacterial morphology deals with size, shape, and arrangement of bacterial cells.

Size of Bacteria

Bacteria are microscopic organisms that are less than 3 micrometeres (μm) in size.Size of cocci range from 0.5 to 3 μm, and the size of a rod shaped bacteria range from 0.15 to 2 μm (width) to 0.5 to 20 μm (length).

Shape of Bacteria

Coccus or **Cocci** are bacterial cells that are spherical, and resemble tiny balls. Eg. *Streptococcus.*

Bacillus or **Bacilli** are bacterial cells that are rod shaped, and resemble a pill.

Eg. *Bacillus anthracis*

Spiral bacteria have twisted or helical morphology that resembles little cork screws. Eg.Gastritis,

Some Bacteria have Other Shapes Such as

Coccobacilli – Elongated spherical or ovoid form.

Filamentous – Bacilli that occur in long chains or threads.

Fusiform – Bacilli with tapered ends.

Arrangement of Cells

Bacteria are also characterized based on how cocci and bacilli aggregate themselves.

Arrangement of Cocci Cells

Singly: Bacteria that appear as single cell, is just called as cocci

Diplococci: These cells are found in pairs and they are found attached to each other

Streptococcus: These bacteria form long chains and remain attached to each other

Staphylococcus: These bacteria are arranged irregularly in clusters like grapes

Arrangement of Bacilli

Singly: Bacteria that exists as single cell, called bacilli

Diplobacilli: These bacteria has two rod shaped cells which are attached to each other

Streptobacilli: Cells are arranged as long chains in these bacteria

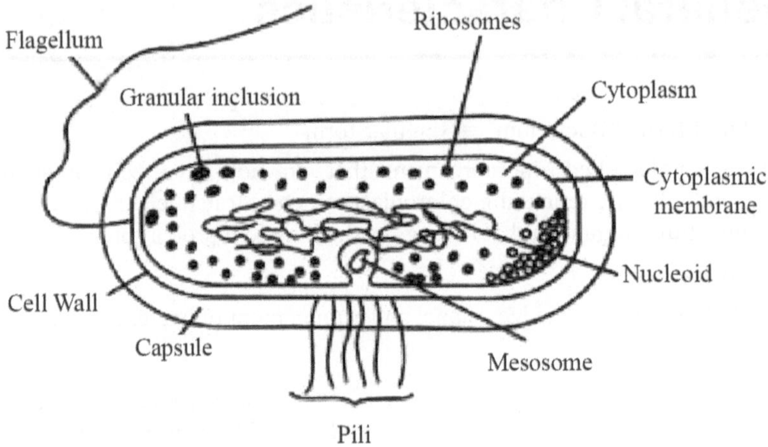

Morphological Characteristics

- Bacteria are single-celled organisms.
- They lack organelles such as chloroplasts and mitochondria.
- Do not have the true nucleus.
- Instead, their DNA, a double strand that is continuous and circular, is located in a nucleoid.
- The nucleoid is an irregularly shaped region that does not have a nuclear membrane.
- Bacteria also have a cell membrane and a cell wall that is often made of peptidoglycan. Together, the cell membrane and cell wall are referred to as the cell envelope.
- Reproduction occurs through binary fission.
- Most numerous organisms on earth.
- Earliest life forms (fossils date 2.5 billion years old..
- Microscopic prokaryotes (no nucleus non membrane-bound organelles).
- Contain ribosomes.
- Surrounded by protective cell wall containing peptidoglycan (protein-carbohydrate).
- Many are surrounded by a sticky, protective coating of sugars called the capsule or glycocalyx (can attach to other bacteria or host).

- Have small rings of DNA called plasmids.
- May have short, hair like projections called pili on cell wall to attach to host or another bacteria when transferring genetic material.
- Most are unicellular.
- Found in most habitats.
- Most bacteria grow best at a pH of 6.5 to 7.0.
- Main decomposers of dead organisms so recycle nutrients.
- Some bacteria breakdown chemical and oil spills.
- Some cause disease.
- Move by flagella.
- Some can form protective endospores around the DNA when conditions become unfavorable; may stay inactive several years and then re-activatewhen conditions favorable.
- Classified by their structure, motility (ability to move), molecular composition, and reaction to stains (Gram stain)
- Most are heterotrophic (can't make their own food..
- Can be aerobic (require oxygen) or anaerobic (don't need oxygen).
- Can be identified by Gram staining (gram positive or gram negative).

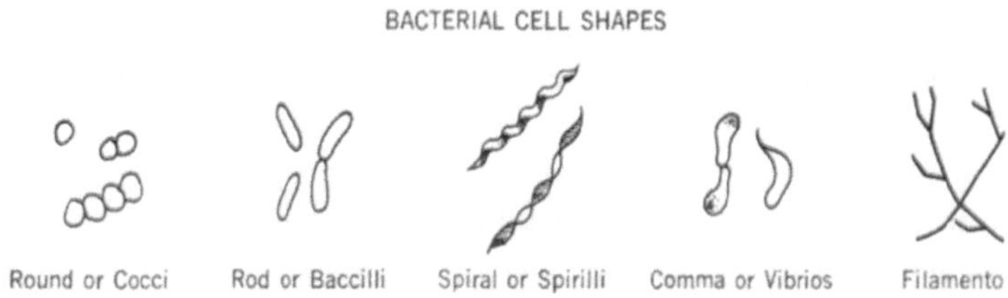

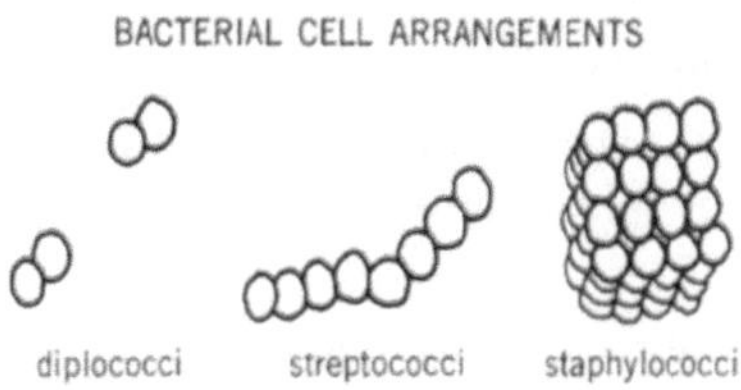

Reproduction

- Most bacteria reproduce asexually by binary fission (chromosome replicates and then the cell divides), budding, fragmentation.
- Bacteria replicate (double in number) every 20 minutes under ideal conditions.
- Bacteria recombine genetic material in 3 ways –conjugation, transformation and transduction.

Conjugation

- Sexual reproductive method.
- Two bacteria form a conjugation bridge or tube between them.
- Pili hold the bacteria together.
- DNA is transferred from one bacteria to the other.

Transformation

- Bacteria pick up pieces of DNA from other dead bacterial cells.
- New bacterium is genetically different from original.

Transduction

- A bacteriophages (virus) carries a piece of DNA from one bacteria toanother.

Important Characteristics of Phytopathogenic Bacteria (PPB)

Shape: Straight to curved rods with rigid cell walls (except filamentous bacteria. Some bacteria assume irregular shapes like " Y, L" Example- *Corynebacterium* – 'V' shape *Agrobacterium* and *Erwinia* - 'L' shape

Decomposition: decomposition is mostly aerobic or oxidative Example- Erwinia (facultative anaerobe)

PPB can be cultured on artificial media. However, pathogenic bacteria grow slowly compared to saprophytes. Majority are flagellate

PPB can be identified based on flagellation, carbohydrate metabolism and pigment production. These are passive invaders i.e. enter into the plants through wounds or natural openings.

Survival – The bacteria survive inside or outside of seed, and in plant debris and spreads by means of water, rain, insects and agricultural implements.

All are susceptible to phages. (Phages are virus that kills bacteria.

All are non-spore forms except Bacillus Sp.

Cell wall rigid and made up of chitin or cellulose

Dispersal- depend on outside agents. They do not spread on the wind as many fungi do. They disseminated by means of splashing water. They are transmitted by insect vectors, nematode and human.

Penetration- cannot penetrate the cuticle of plants but enter the plant through a wound or natural opening.

Classification of Bacteria Bacteria can be classified based on the following characterstics:

1. **Based on shape-** Coccus, Bacillus, Spirillum, Vibrio
2. **Based on staining characters-** Gram positive bacteria, Gram negative bacteria
3. **Based on arrangement of flagella-** Atrichous bacteria, Monotrichous bacteria, Amphitrichous bacteria, Lopotrichous bacteria, Peritrichous bacteria.
4. **Based on nutrition requirement-** Autotrophs, Heterotrophs, Chemoautotrophs, Saprophytic bacteria, Symbiotic bacteria, Pathogenic bacteria

5. **Based on oxygen requirement-** Aerobes, Anaerobes
6. **Based on temperature dependence-** Thermophilic, Mesophiles, psychrotrophs, psychrophiles
7. **Based on DNA relatedness** (phylogenic relationship)

Classification

Kingdom: Prokaryotae

Bacteria: Have cell membrane and cell wall

Division I: Gracilicutes Gram negative bacteria.

Class: Proteobacteria- Mostly single celled bacteria

Family: Enterobacteriaceae

Genus: Erwinia

Family: Pseudomonadaceae

Genus: *Acidovorax, Pseudomonas, Rhizobacter, Rhizomonas,Xanthomonas*

Family: Rhizobiaceae

Genus: *Agrobacterium, Rhizobium*

Division-II: Firmicutes Gram positive bacteria.

Class: Firmibacteria- Mostly single celled bacteria

Genus: *Bacillus, Clostridium*

Class: Thallobacteria Branching bacteria.

Genus: *Arthobacter, Clavibacter, Curtobacterium, Rhodococcus,Streptomyces*

Mollicutes: Mycoplasma like organism- MLO/Phytoplasma/Spiroplasma. Haveonly cell membrane and lacks cell wall.

Division-III: Tenericutes

Class: Mollicutes

Family: Spiroplasmataceae

Genus: *Spiroplasma*

Genus: Undefined, Known as Phytoplasma (MLO)

Fastidious Vascular Bacteria

Fastidious vascular bacteria are similar to the phytopathogenic bacteria but they are introduced directly into the sugar-rich phloem sieve tubes or into the water transporting xylem elements by vascular-feeding insects.The fastidious vascular bacteria may be Fastidious phloem limited bacteria and Fastidious Xylem Limited bacteria (XLB). Fastidious bacteria are bacteria that require specific conditions and substances to grow. They are difficult to grow in laboratories.

The fastidious Phloem limited bacteria are very small bacilli and generally possess Gram- negative prokaryotic cell morphology. Fastidious xylem limited bacteria are walled bacterial pathogens those inhabit the water transporting cells of the xylem of their host plants and are transmitted by xylem feeding sharpshooters and spittlebugs, members of the leafhopper family.

Example- Citrus greening disease, Cucurbit yellow vine, Almond leaf scorch, Alfalfa, Phony Peach, Ratoon stunting, Plum leaf scald.

Major Genera of Plant Pathogenic Bacteria

Genus	Types of disease caused	Other particular features
Gram-negative		
Agrobacterium	Tumors, root Proliferation	Disease uniquely involves transformation of plant host by incorporation of bacterial DNA
Burkholderia	Rot, blight	*Burkholderia glumae* is a rice pathogen of emerging importance
Dickeya	Soft rot	*Dickeya dadantii*, formerly classified as *Erwinia chrysanthemi*, produces copious plant tissue-macerating enzymes
Erwinia	Wilt, necrosis	*Erwinia amylovora* is the causal agent of fireblight
Pantoea	Wilt, necrosis	*Pantoea stewartii* was formerly classified as an *Erwinia* species
Pectobacterium	Soft rot	*Pectobacterium carotovorum* and *Pectobacterium atrosepticum* were formerly classified as *Erwinia* species; produce copious plant tissue–macerating enzymes
Pseudomonas	Leaf spot, wilt, canker, galls	Several species. One species *Pseudomonas syringae* is subdivided into over 40 pathovars
Ralstonia	Wilt	*Ralstonia solanacearum* (formerly *Pseudomonas solanacearum*) is an extremely important plant pathogen
Xanthomonas	Wilt, leaf spot, rot, canker	In earlier classifications, *Xanthomonas campestris* was subdivided into over 120 pathovars. Currently classified as several species, some with pathovar subdivisions
Xylella	Wilt, chlorosis	Only one species *Xylella fastidiosa* is so far defined, with subspecies divisions
Gram-positive		
Clavibacter	Wilt, canker, leaf spot, stunt	*Clavibacter*, *Curtobacterium*, and *Rhodococcus* were formerly classified as *Corynebacterium*
Curtobacterium	Wilt, leaf spot, stem spot	
Rhodococcus	Galls, fasciation	Bacteria produce plant growth hormones
Streptomyces	Rot, common scab	Actinomycetes are a minor group of plant pathogens
Mollicutes		
Spiroplasma	Stunt, wilt	No cell wall, spiral in form, can be cultured
Phytoplasma	Phyllody, stunt	No cell wall, unculturable

Types of disease incited by plant pathogenic bacteria

Disease type	Typical symptoms	Typical causal organisms	Modes of entry into plant
Leaf spot	Small localized lesions which may be chlorotic, necrotic, or water-soaked	*Pseudomonas syringae* pathovars and *Xanthomonas* spp.	Usually through stomata, sometimes wounds
Wilt	Vascular pathogens interfere with water flow, inducing wilting and necrosis	*Ralstonia solanacearum, Xanthomonas campestris* pv. *campestris, Clavibacter* and *Curtobacterium* spp., *Xylella fastidiosa*	Usually through wounds. *X. campestris* pv. *campestris* can enter via hydathodes at leaf margins. *Xylella fastidiosa*, which inhabits xylem, is transmitted by insects
Soft rot	Extensive maceration of plant tissue	*Dickeya Pectobacterium carotovorum*, some *Pseudomonas* spp.	Lenticels and wounds
Cankers and dieback	Tissue necrosis extending from initi infection site. Gives rise to canker lesions or dieback from branches or shoots	*Erwinia amylovora, Xanthomonas axonopodis* pv. *citri, P. syringae* pv. *syringae*	Wounds and stomata (e.g., for *P. syringae*), nectaries (e.g., for *Erwinia amylovora.*
Tumors, galls, and other abnormal plant development	Abnormal plant cell division and (hypertrophy). Phyllody (replacement of floral parts by leaflike structures) and adventitious root formation	*Agrobacterium tumefacien's Pseudomonas savastanoi*, *Rhodococcus fascians* cause galls and tumors. Many phytoplasmas cause phyllody. *Agrobacterium rhizogenes* causes hairy roots	Mostly through wounds. Phytoplasmas, which inhabit phloem, are transmitted via insects.

1. Which method is most preferred for classification of Bacteria?

Sequence analysis of highly conserved region of the bacterial genome such as the small subunit RNA gene is the most preferred method for bacterial classification

2. Which Hierarchical system is followed for bacterial classification?

It is based on hierarchical system like- Domain Bacteria., Phylum, Class, Order, Family, Genus and Species

3. Which book is regarded as reference book for bacterial classification?

Bergey's Manual of Systematic Bacteriology is regarded as reference book for bacterial classification

4. How many volumes do the 1st edition of Bergey's Manual of Systematic Bacteriology have?

The 1st edition published by Williams & Wilkins, Baltimore MD in 4 volumes (1984-89)

5. How many volumes do the 2nd edition of Bergey's Manual of Systematic Bacteriology have?

The current i.e. 2nd edition is edited by Garrity, George M., Brenner, Don J. and Krieg, Noel R. It contains 5 volumes that encompass 6466 species, 2 phyla, 41 classes, 88 orders, 240 families and 1194 genera.

6. What are Taxospecies?

It refers to a group of strains that have a number of phenotypic characteristics in common and can be grouped into distinct Phenotypic group. The named group is called as taxon.

7. What is Nomen-species?

It's a species that was suggested in accordance with the rule of nomenclature, in respective of the inappropriateness to a practical nomenclature system.

8. What are molecular species?

It is a term applied to that group of strains which has high degree of nucleic acid homology. It corresponds to the species in phylogenetic system.

9. What are Genospecies?

It is group of strains that can accomplish genetic exchange in some or the other way.

10. What are the four categories of classification of bacteria according to Bergey's Manual of Systematic Bacteriology (1994)?

1) Major Category 1- Gram negative eubacteria that have cell walls
2) Major Category 2- Gram positive eubacteria that have cell walls
3) Major Category 3- Eubacteria lacking cell walls
4) Major Category 4- The Archaeabacteria

11. What are the characteristic cellular shapes and sizes of *Pseudomonas* bacteria?

Pseudomonas bacteria typically have single cells that appear as straight or curved rods under a microscope. They are not helical in shape and generally measure between 0.5 to 1.5 to 4 microns in size.

12. How do *Pseudomonas* bacteria move?

Pseudomonas bacteria are motile, and they achieve this mobility through the presence of single polar flagellum, the condition of having single flagellum at one end of a bacterium called as monotrichus.These flagella are usually found at one or more ends of the bacterial cell.

13. Are *Pseudomonas* bacteria capable of forming spores?

No, *Pseudomonas* bacteria are non-spore-forming. They do not have the ability to produce spores as some other bacterial species do.

14. What is the Gram staining result for *Pseudomonas* bacteria?

Pseudomonas bacteria are Gram-negative, which means that they do not retain the crystal violet stain during the Gram staining process. Instead, they take up the counter stain and appear pink or red under a microscope.

15. Does *Pseudomonas* cause any disease in plants?

Yes, *Pseudomonas solanacearum* causes brown rot or wilt of potatoes and *Pseudomonas rubrillineans* causes red stripe of sugarcane *Pseudomonas syringae* causes gummosis in *Prunus* spp ..

16. What are the typical cell characteristics of *Xanthomonas*?

Xanthomonas cells are single, straight rods, usually measuring approximately 0.4 by 1.0 microns in size. They are Gram-negative.

17. What is a distinctive feature of *Xanthomonas* in terms of extracellular substances?

Xanthomonas bacteria are known for producing copious extracellular slime.

18. What are the key characteristics of *Xanthomonas* bacteria?

Xanthomonas bacteria are Gram-negative, single straight rods measuring about 0.4 by 1.0 microns. They are motile with a single polar flagellum, non-spore-forming, produce abundant extracellular slime, are strict aerobes with respiratory metabolism, oxidase negative, and their colonies are yellow due to the Xanthomonadin pigment production.

19. Name three important plant diseases caused by *Xanthomonas* bacteria?

X. campestris pv. *citri* causes citrus canker

X. campestris pv. *malvacearum* causes angular leaf sport and black arm of cotton

X. campestris pv. *oryzae* causes bacterial leaf blight of rice

20. What is the typical shape and size of *Agrobacterium* bacteria?

Agrobacterium bacteria are rod-shaped, measuring approximately 0.8 microns in width and 15 microns in length.

21. How do Agrobacterium bacteria achieve motility, and what is a notable feature related to their growth on carbohydrate-containing media?

Agrobacterium bacteria are motile, primarily use peritrichous flagella, of which they typically have 1 to 4. When only one flagellum is present, it is more often lateral than polar. Additionally, when grown on carbohydrate-containing media, these bacteria produce abundant slimy polysaccharide.

22. Name important plant diseases caused by Agrobacterium bacteria?

- *Agrobacterium tumefaciens* causes crown gall of pome and stone fruit trees, brambles and grapes.
- *A. rubi* causes gall of raspberries and black berries.
- *A. rhizogenes* causes hairy root of apple

23. What are the typical characteristics of Erwinia cells in terms of shape, size, and motility?

Erwinia cells are primarily single, straight rods, measuring approximately 0.5 to 1.0 microns in width and 1.0 to 3.0 microns in length. They are motile, except for certain species like *E. stewartii* and *E. dissolvens*, which lack motility.

24. How do Erwinia bacteria grow on artificial media, and what is a notable metabolic product they produce?

Erwinia bacteria are versatile in their growth, thriving both aerobically and anaerobically on artificial media. They are known to produce acid during their metabolic processes.

25. What are the three major groups of species within the genus Erwinia, and what types of diseases or symptoms do they typically cause in plants?

The genus Erwinia consists of three major groups of species

i) Pectolytic bacteria that cause soft rot.

ii) Those that do not cause soft rot but instead lead to dry necrosis and wilt.

iii) Epiphytes that cause neither soft rot nor necrosis.

26. Which genera of bacteria are the most prevalent among plant pathogenic bacteria?

About half of all known plant pathogenic bacteria belong to the genus *Pseudomonas*, about one third to the genus *Xanthomonas*, and about an eighth to the genus *Erwinia.*

27. Enlist important plant diseases caused by *Erwinia*?

i. *E. amylovora* causes fire blight of apple.

ii. *E.tracheiphia* causes vascular wilt of cucurbits.

iii. *E.carotovora* pv. atroseptica causes black leg of potato.

vi *E. carotovora* pv. carotovora causes soft rot of numerous fleshy fruits, vegetables and ornamentals.

28. What are pathovars, and how are they related to subspecies of plant pathogenic bacteria?

Pathovars, abbreviated as "pv.," refer to subspecies or groups of strains of plant pathogenic bacteria that can infect only plants within a certain genus or species. They are a way to classify and differentiate these bacteria based on their host specificity.

29. Classify the major groups of pathogenic species within the genus *Erwinia*, and what types of diseases do these groups cause?

The major groups of pathogenic species within the genus Erwinia have two categories (i) the Amylovora group, whose members do not produce pectic enzymes and cause necrotic or wilt diseases, and (ii) the Carotovora group, whose members have strong pectolytic activity and cause soft rots in plants.

30. Enlist important phytopathogenic bacteria of Group I or Division I?

Bacteria belonging to this group include species of genera

Acetobacter, Acidovorax, Agrobacterium, Bradyrhizobium, Burkholderia, Gluconobacter, Herbaspirillum, Pseudomonas, Ralstonia, Rhizomonas, Xanthomonas, Xylella, Xylophilus.

31. Write economic importance of Phytopathogenic Group I bacteria?

Genus	Economic Importance
Acetobacter	*Acetobacter aceti and Acetobacter diazotrophicus* causes rot and pink disease in pineapple
Agrobacterium	Causes crown gall disease and incites tumor formation in roots, crown and stem etc. Its main function is established in gene transfer due to presence of Ti-plasmid and Ri-plasmid.
Bradyrhizobium	Symbiotic nitrogen fixation with suitable host
Gluconobacter	Causes discoloration in apples, pears and pineapples
Pseudomonas	Different species are pathogenic having wide host range including species of many *Solanaceae, Leguminosae* and *Malvaceae* plants *P. solanacearum* causes brown rot of potato, *P. mangifera* causes leaf spot of mango. *P. syringae* have been divided into many pathovars over 40 that include leaf spot pathogens of soybean and field bean
Ralstonia	*R. solanacearum* causes wilt on solanaceous crops including potato, tabacco, tomato and other plants like peanut and banana
Rhizomonas	*R. suberifaciens* causes bacterial gall of carrot
Xanthomonas	*Xanthomonas* are parasitic on different plants of economic importance. Inoculum may be present in soil or plant debris and seed borne.
Xylophilus	Bacterial necrosis and canker in grapes

32. Enlist important phytopathogenic bacteria of Group II?

Group II are facultative anaerobic Gram-negative rods. This group include different genus *Enterobacter, Erwinia, Pantoea, Pectobacterium, Serratia, Burkholderia*

33. Write economic importance of Phytopathogenic Group II bacteria.

Genus	Economic Importance
Enterobacter	*Enterobacter cloacae* causes brown discoloration of papaya fruits
Erwinia	*Erwinia carotovora* and *Erwinia aroldeae* are known to cause soft rot of potato. *Erwinia carotovora* also causes soft rot of carrot.
Pantoea	*Pantoea stewarti* causes Stewart's wilt of sweet corn
Pectobacterium	*Pectobacterium chrysenthemi,P. caratovorum* cause soft rot in numerous plants
Serratia	*Serratia marcesens* cause yellow vine in cucurbits
Dickeya	*Dickeya dadantii* formely classified as *Erwinia chrysantnemi, causes soft rot of chrysanthemum produces copious plant tissues-macerating enzymes*
Xanthomonas	*Xanthomonas campestris* causes wilt, leaf spot , rot, canker like diseases and subdivided into 120 pathovars and classified recently into several species, some with pathovar subdivisions

Genus	Economic Importance
Xylella	Only one species, *Xylella fastidiosa* is defined ,with subspecies divisions.

35. Gram negative bacteria and diseases caused in plants by them

Burkholderia	Bacterial panicle blight of rice Caused by *Burkholderia glumae*, Onion bulb rot caused by *Burkholderia cepacia*
Agrobacterium	Crown gall and hairy root caused by *Agrobacterium tumefaciens* and *A. rhizogenes,* respectively
Ralsonia	Bacterial wilt of tomato and brown rot of potato caused *by Ralstonia solanacearum*

37. How are phytopathogenic bacteria classified based on host range?

They can be classified as host-specific or non-host-specific based on the range of plants they infect

38. How did Bergey classify plant pathogenic bacteria within his system?

Bergey divided plant pathogenic bacteria in 4 divisions and 7 classes--

Kingdom- Prokaryote

Divison-I-Gracilicutes (Gram negative bacteria that have cell walls)

Class-I- Scotobacteria

Class-II- Anoxyphytobacteria

Class-III- Oxyphytobacteria

Division-II- Firmicutes (Gram positive bacteria that have cell walls)

Class-I- Firmibacteria

Class-II- Thallobacteria

Division-III-Tenricutes (Eubacteria Lacking cell walls)

Class-I-Mollicutes

Division-IV-Mendosicutes (Bcateria with abnormal cell wall)

Class-I- Archbacteria

39. How are phytopathogenic bacteria classified based on their mode of infection?

They can be classified as vascular or non-vascular pathogens depending on whether they infect plant vascular tissues

40. What are the key differences between solid and liquid media used for culturing bacteria?

Solid media, like agar plates, provide a surface for colony growth, making it easy to isolate and study individual bacterial types. Liquid media, such as broths, are used for growing large volumes of bacteria for various applications.

41. How do selective and differential media aid in bacterial culture?

Selective media contain substances that inhibit the growth of certain bacteria while allowing others to thrive, aiding in isolating specific types. Differential media contain indicators that visually distinguish between different bacteria based on their metabolic properties.

42. How do phytopathogenic bacteria enter plant tissues?

They can enter through wounds, natural openings, or by using specialized mechanisms like type III secretion systems

43. What are the key diagnostic methods for identifying phytopathogenic bacteria?

Diagnostic methods include PCR, serological tests, and isolation on selective media

44. How does the genus *Pseudomonas* contribute to plant diseases?

Pseudomonas syringae, for example, causes diseases like bacterial speck in tomato plants

45. How are phytopathogenic bacteria classified based on the disease they cause?

They can be categorized as foliar, vascular, or root pathogens based on the affected plant parts.

46. What are the key characteristics of the genus Erwinia in plant pathology?

Erwinia species are responsible for soft rot diseases in various plants

47. What are the common diseases caused by phytopathogenic bacteria in the Solanaceae family?

Phytopathogenic bacteria in the Solanaceae family can cause diseases like bacterial wilt and bacterial spot

48. What is the role of Bio informatics in the classification of phytopathogenic bacteria?

Bio informatics helps analyze bacterial genomes, identify virulence factors, and study genetic diversity

49. What are some examples of host-specific phytopathogenic bacteria?

Examples include *Pseudomonas syringae* pv. *tomato*, which causes bacterial speck in tomatoes, and *Xanthomonas oryzae* pv. *oryzae*, causing bacterial leaf blight in rice

50. How do phytopathogenic bacteria acquire resistance to copper-based bactericides?

They can develop copper-resistant mechanisms, such as efflux pumps or detoxification enzymes

51. Enlist classes of Phylum Proteobacteria?

The Phylum Proteobacteria are divided into five classes (based on rRNA sequence)

- Alphaproteobacteria
- Betaproteobacteria
- Gammaproteobacteria
- Deltaproteobacteria
- Epsilonproteobacter

52. Differentiate between genera *Pseudomonas* and *Xanthomonas*

	Pseudomonas	*Xanthomonas*
1	The cells are straight or slightly curved with average 0.5-1.0×1.5-5.0μm in size	The cells are rod-shaped, non-spore forming and non-capsulated with average 1.0-2×0.5-0.9 μm in size
2	They have one or several polar flagella	Monotrichously flagellated
3	Appear as whitish or yellow colonies	Usually produce yellow water-soluble pigment in culture
4	*Pseudomonas* causes reddish brown spots that may cause the leaf to distort	*Xanthomonas* causes small brown angular to circular spots with yellow halos

53. How do enzymes contribute to the pathogenicity of bacteria?

Enzymes secreted by phytopathogenic bacteria act as virulence factors, enabling the bacteria to invade plant tissues, degrade cell walls, and suppress the plant's defense mechanisms. This facilitates colonization and nutrient acquisition, leading to disease development.

54. What are some common types of enzymes produced by phytopathogenic bacteria?

Some common types of enzymes include

- **Pectinases** These enzymes break down pectin, a major component of plant cell walls, causing tissue maceration and softening, as seen in soft rot diseases.
- **Cellulases** These enzymes degrade cellulose, another key component of cell walls, leading to tissue disintegration and weakening of plant structures.
- **Proteases** These enzymes break down proteins, disrupting essential cellular processes and potentially inactivating plant defense proteins.
- **Amylases** These enzymes break down starch, a major carbohydrate storage molecule in plants, providing a source of nutrients for the bacteria.

55. Can enzymes produced by phytopathogenic bacteria be used for disease diagnosis?

Yes, the presence of specific enzymes can be used as biomarkers for the detection and identification of phytopathogenic bacteria. Enzyme assays and molecular techniques can be employed to diagnose bacterial diseases in plants.

56. What is the role of the hrp (hypersensitive response and pathogenicity) gene cluster in plant pathogenic bacteria, and how does it relate to the delivery of harpins and Avr proteins?

The role is to enable bacteria to colonize host effectively and induce disease as their genes encode components of type III secretion systems which is involved in the delivery of bacterial effectors, harpins and Avr proteins into host cells.

57. Harpins are known to elicit the hypersensitive response (HR) in plants, a form of programmed cell death. What additional roles, beyond HR elicitation, have been proposed for harpins in the infection process?

Additional roles are their involvement in bacterial attachment, manipulation of plant defense responses, nutrient acquisition, or even suppression of plant immunity.

58. Avr (avirulence) proteins are recognized by corresponding resistance (R) proteins in plants, leading to effector-triggered immunity (ETI). How do plant pathogenic bacteria utilize Avr proteins to their advantage despite their potential to trigger plant defences?

The Avr proteins are utilized by PPB to manipulate host cell and promote infection by suppressing basal defense, alter cellularprocesses for the pathogen's benefit.

Key Points about Avr Proteins

- **Secretion** Avr proteins are secreted by the bacteria into the plant cell using the Type III Secretion System (T3SS), a specialized delivery mechanism.
- **Recognition** Inside the plant cell, Avr proteins are often recognized by corresponding resistance (R) proteins encoded by the plant's genome. This recognition follows a gene-for-gene model.
- **Triggering Defense** The recognition of an Avr protein by an R protein triggers a strong plant defense response known as Effector-Triggered Immunity (ETI). This response often involves a localized cell death called the hypersensitive response (HR), which aims to halt the pathogen's spread.

The Paradox

While Avr proteins seem detrimental to bacteria because they trigger plant defenses, they can also be beneficial in certain situations

- **Suppression of Plant Defenses** Some Avr proteins can suppress basal (general) plant defenses, allowing the bacteria to establish an infection before ETI is activated.
- **Manipulation of Plant Processes** Certain Avr proteins manipulate plant cellular processes to the pathogen's advantage, for example, by altering hormone signaling or nutrient transport.

Evolutionary Arms Race

The interaction between Avr proteins and R proteins represents an ongoing evolutionary arms race. Bacteria evolve new Avr proteins to evade plant recognition, while plants evolve new R proteins to detect these Avr proteins.

Secretion systems in phytopathogenic bacteria

Bacteria utilize various secretion systems to transport proteins across their cell membranes. These systems are essential for nutrient acquisition, communication, and in the case of pathogenic bacteria, delivering virulence factors to host cells.

Here's a summary of the main bacterial secretion systems

Sec-dependent Systems

- **Sec (General Secretory Pathway)** The most common system, transporting unfolded proteins across the inner membrane.

- **Tat (Twin Arginine Translocation)** Transports folded proteins across the inner membrane.

Gram-Negative Bacteria Specific Systems

- **Type I (ABC Transporter)** A one-step system that secretes proteins directly from the cytoplasm to the extracellular environment.
- **Type II (Two-Step)** First transports proteins to the periplasm, then across the outer membrane.
- **Type III (Injectisome)** A needle-like apparatus that injects proteins directly from the bacterial cytoplasm into host cells. Crucial for many pathogenic bacteria.
- **Type IV (Pilus-like)** Can transport proteins or DNA between bacteria or into host cells.
- **Type V (Autotransporter)** Proteins transport themselves across the outer membrane after being transported to the periplasm by the Sec system.
- **Type VI (T6SS)** A contractile system that can deliver proteins to both prokaryotic and eukaryotic cells.

Gram-Positive Bacteria Specific System

- **Type VII (T7SS)** A specialized system found in some gram-positive bacteria, particularly those associated with the human microbiome.

Additional Notes

- The systems listed above are primarily for gram-negative bacteria, which have an inner and outer membrane. Gram-positive bacteria have a single membrane, and their secretion systems are often variations of the Sec and Tat pathways.
- Some secretion systems are conserved across many bacterial species, while others are found only in specific groups of bacteria.
- The study of bacterial secretion systems is an active research area, and new systems and functions are still being discovered.

3

Virus and Their General Characteristics

Plant viruses are unusual group of plant pathogens that almost depend on host cellular machinery for their survival, hence categorized under obligate intracellular parasites. They are smart entities as they encode very few proteins but take the advantage of host cellular metabolism to complete their replication and movement within plants cells. These viruses passively enter the host cells through various means such as wounds and injury, environmental factors, or by several vectors. Genomic material, either RNA or DNA of viruses, disassembles inside the plant cells, and replicates and transcribes mRNA to protein with the help of host factors and energy. Once they manage to assemble their required genetic materials and capsid protein, virions are then released to neighboring cells for further infecting other cells. They even reach to distant cells through plant vascular system causing systemic infection and different symptoms in host plant.

Plant Viruses - A Deeper Dive with Examples

- **Obligate Parasites:** Plant viruses are not capable of independent life. They absolutely require a living host plant cell to reproduce and carry out their life cycle. They cannot replicate outside of a cell.
- **Infect a Wide Range of Plants:** Different viruses specialize in infecting different plant species. Some have a narrow host range, only infecting a few related plants, while others have a broader host range. This specificity is often determined by the interactions between viral proteins and host cell receptors.

  Example: *Tobacco mosaic virus (TMV)* primarily infects tobacco plants and other members of the *Solanaceae* family (like tomatoes and peppers).
- **Impact on Agriculture and Horticulture:** Viral diseases can severely impact crop yields, reduce the quality of produce, and even kill plants. This leads to significant economic losses for farmers and the agricultural industry. In horticulture, viral infections can ruin the aesthetic value of ornamental plants.

  Example: *Tomato spotted wilt virus (TSWV)* can cause significant losses in tomato, peanut, and tobacco crops.

Structure - The Building Blocks of a Virus

- **Genome**
- **Single-stranded RNA (ssRNA):** The most common type of plant virus genome. ssRNA can be positive-sense (like mRNA, ready for translation) or negative-sense (must be transcribed into a positive-sense strand first).

- **Example:** *Tobacco mosaic virus (TMV)* has a positive-sense ssRNA genome.
- **Double-stranded RNA (dsRNA):** Less common, but still significant. dsRNA viruses often have complex replication strategies.
- **Example:** *Rice dwarf virus* is a dsRNA virus that affects rice plants.
- **Single-stranded DNA (ssDNA) and Double-stranded DNA (dsDNA):** These are less frequent in plant viruses compared to RNA viruses.
- **Example:** *Bean golden mosaic virus* is an ssDNA virus that infects beans.

- **Capsid - The Protective Shell**
  - **Protein Coat:** The capsid is composed of many identical protein subunits called capsomeres. These proteins assemble in a highly organized way to form the capsid.
  - **Protection:** The capsid protects the viral genome from degradation by enzymes (like nucleases) in the environment or within the plant cell.
  - **Cell Entry:** The capsid plays a role in attaching to the host cell and facilitating entry into the cell's interior. Specific proteins on the capsid surface recognize and bind to receptors on the plant cell membrane.
  - **Shapes**
    - **Rod-shaped:** Elongated, cylindrical structures. Tobacco Mosaic Virus (TMV) is a classic example.
    - **Spherical (Isometric):** Roughly spherical or polyhedral.
    - **Example:** *Cucumber mosaic virus (CMV)* has an isometric capsid.
    - **Complex:** Some viruses have more intricate structures, sometimes with multiple layers of protein or other appendages.
  - **Envelope - A Rare Outer Layer**
  - **Usually Absent:** Unlike many animal viruses, plant viruses typically lack a lipid envelope.
  - **Acquired from Host:** When present, the envelope is not made by the virus itself. Instead, it is derived from the host plant cell's membrane as the virus particle buds out of the cell.
    - **Example:** *Tomato spotted wilt virus (TSWV)* is an example of a plant virus that has an envelope.

Function - How Viruses Work

- **Replication - Hijacking the Host**
- **Host Machinery:** Viruses are masters of deception. They inject their genetic material into the plant cell and then use the cell's own ribosomes, enzymes, and other cellular machinery to make copies of their genome and produce viral proteins.
- **Viral Proteins:** These proteins include capsid proteins (for new virus particles), replication enzymes, and proteins that interfere with the plant's defense systems.
- **Assembly:** Newly synthesized viral RNA or DNA and capsid proteins assemble into new virus particles.

- **Disease Induction - Making the Plant Sick**
  - **Disruption of Processes:** Viruses can disrupt various essential plant cell processes, including photosynthesis, nutrient transport, and cell signaling.
  - **Symptoms:** The specific symptoms caused by a virus depend on the virus, the host plant, and environmental conditions. Common symptoms include:
    - **Mosaic Patterns:** Patches of light and dark green on leaves.
    - **Example:** *Tobacco mosaic virus* causes a characteristic mosaic pattern on tobacco leaves.
    - **Leaf Curling or Distortion:** Abnormal leaf shape.
    - **Example:** *Tomato yellow leaf curl virus* causes severe leaf curling and distortion in tomato plants.
    - **Stunting:** Reduced growth.
    - **Example:** Many viruses can cause stunting in infected plants, reducing their overall size and yield.
    - **Necrosis:** Localized tissue death.
    - **Example:** *Potato virus Y (PVY)* can cause necrotic lesions on potato leaves and tubers.
    - **Chlorosis:** Yellowing of leaves due to chlorophyll loss.
    - **Example:** *Barley yellow dwarf virus* causes chlorosis in barley and other cereal crops.
  - **Reduced Yield and Quality:** Viral infections can significantly reduce the yield and quality of crops, making them unmarketable.

Transmission - Spreading the Infection

- **Vectors - The Messengers**
- **Insects:** Aphids, whiteflies, thrips, leafhoppers, and other insects are the most common vectors for plant viruses. They acquire the virus while feeding on an infected plant and then transmit it to a healthy plant when they feed again.
- **Example:** *Aphids* are vectors for many viruses, including *Cucumber mosaic virus*. *Whiteflies* transmit viruses like *Tomato yellow leaf curl virus*.
- **Other Vectors:** Fungi, nematodes (microscopic worms), and mites can also transmit some plant viruses.
  - **Example:** Certain *fungi* can transmit viruses that cause diseases in crops like wheat and barley.
- **Persistent vs. Non-Persistent**
  - **Persistent:** The virus replicates within the vector and can be transmitted for a long period (even for the vector's lifetime).
  - **Example:** *Circoviruses* are examples of viruses that exhibit persistent transmission in their insect vectors.
  - **Non-Persistent:** The virus is acquired by the vector but is lost quickly. Transmission is usually short-lived.

- **Example:** *Aphids* often transmit viruses in a non-persistent manner. They acquire the virus while feeding, but lose the ability to transmit it after a short period.

- **Mechanical Transmission - Accidental Spread**
- **Contaminated Tools:** Pruning shears, knives, and other tools can become contaminated with virus particles and spread the virus when used on healthy plants.
- **Human Contact:** Handling infected plants and then healthy plants can also transmit the virus.
  - **Example:** *Tobacco mosaic virus (TMV)* is easily transmitted mechanically because it is a very stable virus and can survive for long periods on surfaces.
- **Seed Transmission - From Parent to Offspring:**
- **Infected Seeds:** Some viruses can be present within the seed itself, infecting the seedling when it germinates.
  - **Example:** *Bean common mosaic virus* is seed-transmitted in beans.
- **Pollen Transmission - During Pollination**
- **Virus in Pollen:** Viruses can sometimes be carried in pollen grains and transmitted to healthy plants during pollination.
- **Example:** *Prunus necrotic ringspot virus* can be transmitted through pollen in cherry trees.
- **Vegetative Propagation - Cloning with a Virus**
- **Cuttings, Tubers, etc.:** If a plant used for vegetative propagation (like cuttings or tubers) is infected with a virus, the new plants grown from these parts will also be infected.
  - **Example:** Many fruit trees and ornamental plants are propagated vegetatively, and this can be a way to spread viruses if the parent plant is infected.
- **Grafting - Joining Infected and Healthy:**
- **Virus Transfer:** Grafting an infected plant part onto a healthy plant will almost certainly transmit the virus to the healthy plant. This is sometimes used intentionally in research to study virus infections.
- Example Apple chlorotic leaf spot virus, Apple Mosaic Virus.

Plant Virus Types (by Genome)

- **RNA Viruses (Most Common)**
- **Positive-sense ssRNA:** The viral RNA can be directly translated into proteins by the host cell's ribosomes.

  Examples: Tobacco mosaic virus (TMV) (rod-shaped, causes mosaic disease in tobacco)
  - *Cucumber mosaic virus (CMV)* (spherical, infects a wide range of plants)
  - *Tomato spotted wilt virus (TSWV)* (spherical, has an envelope, affects tomatoes, peanuts, etc.)
  - *Potato virus X (PVX)* (rod-shaped, infects potatoes)

- **Negative-sense ssRNA:** The viral RNA is complementary to the mRNA needed for protein synthesis. The virus must carry an RNA-dependent RNA polymerase to make mRNA.
  - *Examples:*
  - *Lettuce necrotic yellows virus (LNYV)* (bacilliform, affects lettuce)
  - *Tomato black ring virus (TBRV)* (spherical, affects tomatoes and other plants)
- **Double-stranded RNA (dsRNA):** The viral genome consists of two strands of RNA.
  - *Examples:*
  - *Rice dwarf virus (RDV)* (spherical, affects rice)
  - *Maize rough dwarf virus (MRDV)* (spherical, affects maize)
  - *Grapevine fanleaf virus (GFLV)* (spherical, affects grapevines)
- **DNA Viruses (Less Common):**
- **Single-stranded DNA (ssDNA):** The viral genome consists of a single strand of DNA.
- *Examples: Bean golden mosaic virus (BGMV)* (geminivirus, affects beans)
- *Tomato yellow leaf curl virus (TYLCV)* (geminivirus, affects tomatoes)
- **Double-stranded DNA (dsDNA):** The viral genome consists of two strands of DNA.
  - *Examples:*
  - *Cauliflower mosaic virus (CaMV)* (pararetrovirus, affects cauliflower and other cruciferous plants)
  - *Fig mosaic virus (FMV)* (affects fig trees)

Other Classifications (Less Common, but Important)

- **Geminiviruses:** A large group of ssDNA viruses with a characteristic twin (geminate) particle structure. Many are economically important, affecting crops like tomatoes, beans, and cassava. *Example: Tomato yellow leaf curl virus (TYLCV).*
- **Closteroviruses:** A group of positive-sense ssRNA viruses with long, filamentous particles. They often infect fruit trees and grapevines. *Example: Grapevine leafroll-associated virus (GLRaV).*
- **Potyviruses:** A large and economically significant group of positive-sense ssRNA viruses. They cause a wide range of diseases in various crops. *Example: Potato virus Y (PVY).*
- **Begomoviruses:** A genus of geminiviruses that are transmitted by whiteflies and cause significant yield losses in many crops. *Example: Bean golden mosaic virus (BGMV).*

Viruses vs. Viroids

Feature	Virus	Viroid
Structure		
Genetic Material	DNA or RNA (single- or double-stranded)	RNA only (single-stranded, circular)
Protein Coat	Present (called a capsid)	Absent
Envelope	May or may not be present	Absent
Size	Larger (20-300 nanometers)	Much smaller (20-100 nanometers)
Coding Capacity	Can code for proteins (e.g., capsid proteins, replication enzymes)	Does not code for any proteins
Host Range	Can infect bacteria, archaea, fungi, plants, and animals	Primarily infects plants
Replication	Uses host cell machinery to replicate genetic material and produce viral proteins	Uses host cell machinery to replicate its RNA, but does not produce proteins
Disease	Causes a wide range of diseases in various organisms	Causes specific plant diseases, often with characteristic symptoms
Examples	HIV, influenza virus, tobacco mosaic virus	Potato spindle tuber viroid, hop stunt viroid

Key Takeaways

- **Simplicity:** Viroids are much simpler in structure than viruses. They consist only of a short, circular RNA molecule without any protein coat.
- **Coding Capacity:** Viruses have the genetic information to produce proteins, while viroids do not.
- **Host Specificity:** While viruses can infect a wide range of organisms, viroids are known to infect primarily plants.
- Viroids are like naked strands of RNA that can cause disease in plants.

1. What distinguishes a plant virus from other plant pathogens like bacteria or fungi at the molecular level?

Plant viruses lack cellular structures (ribosomes, cell membranes, etc.) and possess a genome of either RNA or DNA enclosed in a protein coat (capsid), unlike bacteria or fungi.

2. How does the positive-sense ssRNA genome of a plant virus like TMV facilitate its replication in the host cell?

The positive-sense ssRNA can be directly translated by the host cell's ribosomes into viral proteins, including the RNA-dependent RNA polymerase needed for genome replication.

3. What is the role of RNA-dependent RNA polymerase (RdRP) in the replication of RNA viruses?

RdRP is an enzyme essential for RNA virus replication, as it synthesizes new RNA strands using RNA as a template, a process not typically carried out by the host cell's enzymes.

4. How do plant viruses exploit the host cell's protein synthesis machinery for their own benefit?

Viruses hijack the host cell's ribosomes, tRNA, and other translation factors to synthesize viral proteins, including those required for replication, capsid formation, and movement within the plant.

5. What are the different mechanisms by which plant viruses move within the host plant?

Plant viruses can move cell-to-cell through plasmodesmata, often aided by viral movement proteins, and systemically through the plant's vascular system (phloem).

6. How do viral movement proteins (MPs) facilitate the cell-to-cell movement of plant viruses?

MPs modify plasmodesmata, the small channels connecting plant cells, allowing the passage of viral particles or their components.

7. What is the significance of the icosahedral or rod-shaped structure of viral capsids?

These structures provide stability and protection for the viral genome and often play a role in host cell recognition and entry.

8. How do plant viruses manipulate the host plant's defense responses?

Viruses encode proteins that can suppress or interfere with the plant's RNA silencing machinery, a key antiviral defense mechanism.

9. What is RNA silencing and how does it function as a plant antiviral defense?

RNA silencing involves the production of small interfering RNAs (siRNAs) that target and degrade viral RNA, thus limiting viral replication.

10. How do plant viruses overcome the challenges of systemic movement through the phloem?

Viruses often utilize specific viral proteins or rely on the plant's own transport mechanisms to move through the phloem, which is specialized for long-distance transport of sugars and other molecules.

11. What are the different types of vectors involved in plant virus transmission, and how do they acquire and transmit viruses?

Common vectors include insects (aphids, whiteflies), nematodes, fungi, and mites. They acquire viruses by feeding on infected plants and transmit them to healthy plants during subsequent feeding.

12. What distinguishes persistent and non-persistent transmission of plant viruses by insect vectors?

Persistent transmission involves the virus replicating within the vector, allowing for prolonged transmission. Non-persistent transmission involves the virus being acquired and transmitted briefly, without replication in the vector.

13. What is the role of plant virus receptors in host specificity?

Specific receptors on the plant cell surface interact with viral proteins, determining whether a virus can successfully infect a particular plant species.

14. How can seed transmission contribute to the spread of plant viruses?

Seed transmission allows viruses to be passed from parent plants to offspring, ensuring infection in the next generation.

15. What are viroids, and how do they differ from plant viruses?

Viroids are small, circular, non-coding RNA molecules that are infectious and replicate autonomously in plant cells, unlike viruses, they lack a protein coat.

16. How do viroids replicate in the absence of protein-coding genes?

Viroids utilize the host cell's RNA polymerase to replicate their RNA genomes through a rolling circle mechanism.

17. What are some examples of plant diseases caused by viroids?

Potato spindle tuber viroid, hop stunt viroid.

18. How does the environment influence plant virus infections and symptom development?

Environmental factors like temperature, humidity, and light can affect both the plant's susceptibility to infection and the severity of symptoms.

19. What are some strategies for managing plant viral diseases in agriculture?

Strategies include planting resistant varieties, vector control, sanitation practices, and using virus-free planting material.

4

Host-Pathogen Interaction and Genetic Basis of Variability in Plant Pathogens

1. Host-Pathogen Recognition

- **Pathogen-Associated Molecular Patterns (PAMPs)** Plants have receptors that can recognize conserved molecules on pathogens, called PAMPs (e.g., bacterial flagellin). This triggers a basal defense response called PAMP-Triggered Immunity (PTI).
- **Effector Proteins** Pathogens secrete effector proteins into plant cells to suppress PTI and facilitate infection. Plants have evolved resistance (R) proteins that can recognize specific effector proteins, leading to a stronger defense response called Effector-Triggered Immunity (ETI).

2. Genesis of Interaction

- **Entry** Pathogens enter plants through natural openings (stomata, hydathodes) or wounds. Some can directly penetrate plant tissues using specialized structures or enzymes.
- **Colonization** Once inside, pathogens multiply and spread within plant tissues, often utilizing enzymes to break down cell walls and obtain nutrients.
- **Manipulation of Host** Pathogens manipulate host physiology through effector proteins, suppressing defense, and altering hormone signaling, and redirecting resources to support their growth.

3. Plant Defense Mechanisms

- **Physical Barriers** The plant's cuticle (waxy layer) and cell walls act as the first line of defense, hindering pathogen entry.
- **Chemical Defenses** Plants produce antimicrobial compounds (phytoalexins) and reactive oxygen species (ROS) to inhibit pathogen growth.
- **Hypersensitive Response (HR)** In ETI, the plant sacrifices infected cells through programmed cell death (HR) to prevent pathogen spread.
- **Systemic Acquired Resistance (SAR)** A successful HR can trigger SAR, a long-lasting, broad-spectrum resistance throughout the plant.
- **RNA Silencing** Plants use RNA interference (RNAi) to target and degrade pathogen genes.

Procedure of plant defense in a Nutshell

1. **Pathogen Contact** The pathogen comes into contact with the plant surface.
2. **Recognition** Plant receptors recognize PAMPs or effector proteins.

3. **Defense Activation** The plant mounts PTI or ETI defense
4. **Pathogen Counter attack** The pathogen attempts to suppress plant defenses using effectors.
5. **Outcome** The outcome depends on the balance between the plant's defense mechanisms and the pathogen's virulence strategies.

Induced Systemic Resistance (ISR) and Systemic Acquired Resistance (SAR) are two important defense mechanisms in plants that help them resist attacks from pathogens and pests. While both ISR and SAR enhance a plant's ability to defend itself, they are activated by different pathways and use different signaling molecules.

Systemic Acquired Resistance (SAR)

SAR is a "whole-plant" resistance response that occurs following an earlier localized exposure to a pathogen. The initial infection triggers a cascade of signaling events, primarily mediated by salicylic acid (SA., that leads to the activation of defense-related genes throughout the plant. This systemic response provides long-lasting protection against a broad spectrum of pathogens.

- **Key features of SAR**
  - Triggered by a local infection
  - Mediated by salicylic acid (SA.
  - Provides long-lasting, broad-spectrum resistance

Induced Systemic Resistance (ISR)

ISR is a form of induced resistance that is triggered by beneficial microbes colonizing the plant's roots. These microbes, such as plant growth-promoting rhizobacteria (PGPR), activate a signaling pathway in the plant, predominantly involving jasmonic acid (JA. and ethylene (ET), which primes the plant's defenses. ISR provides enhanced protection against a wide range of pathogens and pests.

- **Key features of ISR:**
  - Triggered by beneficial root-colonizing microbes
  - Mediated by jasmonic acid (JA. and ethylene (ET)
  - Provides enhanced resistance against a broad spectrum of pathogens and pests

Differences between SAR and ISR

Feature	SAR	ISR
Trigger	Local infection	Beneficial root-colonizing microbes
Signaling molecules	Salicylic acid (SA.	Jasmonic acid (JA. and ethylene (ET)
Specificity	Broad spectrum	Broad spectrum
Duration	Long-lasting	Shorter-lasting

While both SAR and ISR play crucial roles in plant defense, they are complementary rather than mutually exclusive. The activation of one pathway does not necessarily inhibit the other, and in some cases, both pathways may be activated simultaneously, providing the plant with a more robust and comprehensive defense response.

1. What is host-pathogen interaction in plants?

Host-pathogen interaction in plants refers to the dynamic relationship between a plant (the host) and a microorganism (the pathogen) such as fungi, bacteria, viruses, or nematodes, where the pathogen can cause diseases in the host.

2. How do plants recognize pathogens?

Plants recognize pathogens through pattern recognition receptors (PRRs) that detect pathogen-associated molecular patterns (PAMPs) or through resistance (R) proteins that recognize specific pathogen effectors.

3. What are PAMPs and how do they trigger plant defense?

PAMPs are pathogen-associated molecular patterns which are conserved molecular structures in pathogens. When Pattern Recognizing Receptors (PRR) on plant cells detect PAMPs, it triggers an innate immune response, including the production of antimicrobial compounds.

4. What are effectors proteins, and how do they affect plant defense?

Effectors are proteins secreted by pathogens to manipulate the host. Incompatible plant-pathogen interactions involve recognition of effectors by R proteins, leading to a strong defense response.

5. Explain the difference between compatible and incompatible plant-pathogen interactions.

Compatible interactions result in successful pathogen colonization and disease, while incompatible interactions result in the activation of plant defense mechanisms and resistance against the pathogen.

6. What is the hypersensitive response in plant defense?

The hypersensitive response is a localized cell death and defense reaction at the site of pathogen attack. It helps restrict pathogen spread in incompatible interactions.

7. How do plants defend against necrotrophic pathogens?

Plants defend against necrotrophic pathogens by activating defense mechanisms involving the production of toxic compounds and cell death to limit pathogen growth.

8. What are systemic acquired resistance (SAR) and induced systemic resistance (ISR)?

SAR and ISR are plant defense mechanisms that involve the activation of resistance throughout the plant after an initial infection, making the plant more resistant to subsequent pathogen attacks.

9. What role does phytohormone salicylic acid (SA) play in plant defense?

SA is a key signaling molecule in plant defense against biotrophic pathogens. It triggers the expression of defense-related genes and is involved in SAR.

10. How do jasmonic acid (JA. and ethylene (ET) mediate defense against necrotrophic pathogens?

JA and ET are phytohormones that play a crucial role in defense against necrotrophic pathogens, promoting the production of toxic compounds and cell death.

11. What is the role of plant secondary metabolites in defense against pathogens?

Secondary metabolites like alkaloids, terpenoids, and phenolics act as chemical defenses in plants, deterring or inhibiting pathogen growth.

12. How do plant cell walls contribute to defense against pathogens?

Plant cell walls provide physical and chemical barriers against pathogens. They can also be reinforced during infection as part of the defense response.

13. Explain the role of reactive oxygen species (ROS) in plant defense.

ROS are molecules involved in plant defense against pathogens. They can have both toxic and signaling roles, contributing to pathogen cell damage and defense gene activation.

14. What do you understand by RNA interference (RNAi) and its role in plant defense against viruses?

RNAi is a mechanism in which small RNA molecules target and silence viral RNA in plant cells, providing a defense against viral infections.

15. How do plants adapt to long-term pathogen pressure through gene-for-gene interactions?

Gene-for-gene interactions involve the co-evolution of plant resistance (R) genes and pathogen avirulence (Avr) genes, leading to durable resistance in plants.

16. What is effector-triggered immunity (ETI) in plant defense?

ETI is a type of plant immunity triggered by the recognition of specific pathogen effectors by R proteins, leading to a rapid and robust defense response.

17. How do beneficial microbes, such as mycorrhizal fungi and rhizobacteria, influence plant-pathogen interactions?

Beneficial microbes can enhance plant resistance by priming the plant's immune system or competing with pathogens for resources.

18. Explain the concept of gene silencing in plant immunity.

Gene silencing involves the suppression of specific plant genes in response to pathogen attack, either to limit the pathogen's virulence or to modulate plant defense.

19. What is the role of callose deposition in plant defense?

Callose is a polysaccharide that accumulates at the site of pathogen infection, reinforcing the plant cell wall and preventing pathogen penetration.

20. How do temperature and environmental factors influence plant-pathogen Interactions?

Temperature and environmental conditions can affect the outcome of plant-pathogen. Interactions by influencing pathogen growth and the plant's ability to defend itself.

21. What are transgenic plants, and how can they be used to enhance resistance against pathogens?

Transgenic plants are genetically modified to express specific genes, including those for pathogen resistance, which can enhance their ability to resist pathogen attacks.

22. How do plant pathogens overcome host defenses and cause disease?

Pathogens can overcome host defenses through various mechanisms, including effectors diversification, mutation, and the suppression of plant defense.

23. What is the role of RNA-based immunity in plant-pathogen interactions?

RNA-based immunity involves the recognition and degradation of pathogen-derived RNAs in plants, providing another layer of defense.

24. How do non-host plants defend against pathogens that typically infect other plant species?

Non-host resistance involves strong defense responses in plants against pathogens that typically infect other species, limiting pathogen colonization.

25. What are the potential applications of understanding plant-pathogen interactions for agriculture and crop protection?

Understanding plant-pathogen interactions can lead to the development of more effective and sustainable strategies for crop protection, including the breeding of disease-resistant crops and the development of environmentally friendly control methods.

26. What is variability in plant pathogens?

Variability in plant pathogens refers to the ability of these microorganisms to exhibit genetic diversity and adapt to changing environments or host plants.

27. Why is variability important in plant pathogens?

Variability allows plant pathogens to evade host defenses, develop resistance to control measures, and exploit different plant hosts, making them more successful and challenging to manage.

28. How do plant pathogens exhibit variability in terms of host range?

Plant pathogens can adapt to infect a wide range of plant species by developing new pathogenic strains or variants.

29. What are the genetic mechanisms underlying variability in plant pathogens?

Genetic mechanisms include mutation, recombination, and horizontal gene transfer, which can introduce novel traits and diversity into pathogen populations.

30. What role does mutation play in generating variability in plant pathogens?

Mutation introduces genetic changes, which can lead to the emergence of new pathogen strains with altered characteristics, such as increased virulence or resistance.

31. How does recombination contribute to the variability of plant pathogens?

Recombination involves the exchange of genetic material between different pathogen strains, leading to the creation of hybrid strains with unique traits.

32. Can environmental factors influence the variability of plant pathogens?

Yes, environmental conditions can select for specific pathogen variants that are better adapted to particular climates or ecological niches.

33. What is the significance of variability in plant pathogens for plant disease management?

Variability can complicate disease management strategies by reducing the effectiveness of control measures and necessitating ongoing research and adaptation.

34. How do plant pathogens adapt to different environmental conditions through variability?

They can develop variants with specific adaptations to varying environmental factors, such as temperature, humidity, and nutrient availability.

35. What are some examples of variability in plant viruses?

Examples include the emergence of new strains of the Potato virus Y (PVY) that overcome plant resistance and the variability in strains of the Tomato mosaic virus (ToMV) affecting different tomato cultivars.

36. How does variability in plant pathogens affect the development of resistance in crops?

Variability can lead to the breakdown of resistance in crops as pathogens adapt to overcome the plant's defense mechanisms.

37. How can the study of variability in plant pathogens inform breeding programs for disease-resistant crops?

Understanding pathogen variability helps breeders select and develop resistant crop varieties that can withstand diverse pathogen challenges.

38. What are the implications of variability in plant pathogens for the effectiveness of chemical and biological control methods?

Variability may reduce the long-term effectiveness of control methods, as pathogens can evolve resistance to chemicals or antagonistic microorganisms.

39. How do plant pathogens evolve to overcome host resistance mechanisms through variability?

They can acquire new genetic traits that allow them to evade or suppress host resistance genes, resulting in disease breakthroughs.

40. What tools and techniques are used to study and characterize variability in plant pathogens?

Molecular biology techniques like genotyping, sequencing, and Bio informatics are used to analyze pathogen populations and identify genetic variation.

41. Can variability in plant pathogens lead to emerging diseases in agricultural systems?

Yes, variability can lead to the emergence of new diseases when pathogens adapt to infect previously unaffected crops.

42. What is the role of host-pathogen interactions in shaping the variability of plant pathogens?

Host-pathogen interactions drive selection pressures, favouring pathogen variants that can better infect and reproduce within specific host species.

43. How do plant pathogens evolve to evade the plant immune system through variability?

They can alter their surface molecules or effectors proteins to avoid recognition by the plant immune system, allowing them to establish infections.

44. What are the economic and ecological consequences of variability in plant pathogens?

Economically, variability can lead to crop losses, increased production costs, and reduced food security. Ecologically, it can impact natural ecosystems by affecting native plant populations.

45. How does variability in plant pathogens impact global food security?

It can threaten food security by reducing crop yields, increasing the risk of crop failures, and limiting the availability of disease-resistant crop varieties.

46. Are there any examples of beneficial variability in plant pathogens, such as Bio control agents?

Yes, some beneficial microorganisms can exhibit variability to adapt to and control harmful plant pathogens, offering potential bio control solutions.

47. How can farmers and agricultural scientists manage and mitigate the impacts of variability in plant pathogens?

Management strategies include crop rotation, genetic resistance, integrated pest management (IPM), and the development of new control measures.

48. What are the challenges in predicting and responding to changes in the variability of plant pathogens?

Challenges include the unpredictability of pathogen evolution, the need for continuous monitoring, and the development of timely responses to emerging threats.

49. How can advanced molecular techniques help in understanding and managing variability in plant pathogens?

Advanced techniques can provide insights into pathogen genetics, epidemiology, and evolution, facilitating more targeted and effective management strategies.

50. What are the future directions of research in the field of variability in plant pathogens, and how might they benefit agriculture and plant disease management?

Future research may focus on predicting pathogen behavior, developing novel control methods, and improving our understanding of host-pathogen interactions, ultimately enhancing crop protection and food production and security.

5

Plant Disease Epidemiology

Plant disease epidemiology is the study of how plant diseases develop and spread. It involves understanding the causes of diseases, how they spread, and their impact on plants and crops. This knowledge is used to develop strategies for managing and controlling plant diseases, such as predicting outbreaks, breeding resistant plants, and implementing cultural and chemical controls.

Element of an epidemic: Plant disease epidemics develop as a result of the timely combination of the element in plant disease: Susceptible host plant, a virulent pathogen and favourable environmental conditions over a relatively long period of time. Humans may unwillingly help initiate and develop epidemics. Thus, the chance of an epidemic increases when the susceptibility of the host and virulence of the pathogen are greater, as the environmental conditions approach the optimum level for pathogen growth, reproduction and spread and as the duration of allfavourable conditions.

Necessary Conditions for Development Of Epidemic are

- The presence of abundant viable inoculum inoculums early in the season near the host plant.
- Abundant and timely liberation of inoculums.
- The presence of virulent physiological races in the inoculums.
- Rapid and wide spread dissemination of inoculums..
- A concentrated and extensive population of susceptible host in the most susceptible stage when the inoculums are disseminated.
- Optimum moisture and temperature for rapid and abundant germination of inoculums andv entry of the pathogen into the host.
- Optimum weather conditions for rapid growth and abundant production of the pathogen.

Factor Affecting Disease Development

There are several factors which influence the disease development leading to epidemic factors host, pathogen, environment-moisture, temperature, light, soil pH, oxygen and carbon dioxide in soil and time

Host: The attempts have been made to develop varieties resistant to importantv pathogen and their virulent races. Good crop varieties resistant to pathogens and their virulent races have been developed but certain time new races are also found in nature and even varieties resistant races of the pathogens turn susceptible as in the cased UG -99 in wheat rust. Usually the varieties are highly resistant and moderately resistant but their performance depends on the virulence of the races of the pathogen. The degree of genetic uniformity of crop variety towards susceptibility is important in increased disease severity. Collateral host are able to increase the quantity of inoculums.

Pathogen: Fungi, bacteria, viruses, nematodes, phanerogamic plant parasites attack the plant to cause diseases. Some of them reproduce rapidly resulting in enormous inoculum. The inoculums get dispersed through air, water, insects, soil and seed. The quantum of inoculums which gets associated with the host, its virulence, rapid germination and power of causing infection is important and is termed inoculums potential. Fungi reproduce through the production of conidia, sporangia, chlamydospores, spores, sexual bodies; bacteria through vegetative propagation ; viruses in living cells; nematodes through sexual reproduction, phanerogamic plant parasites through seed. The pathogen propagules may be present in the crop field itself or on a host or long distance. The time taken in a plant to get diseased depends upon this distance.

Environmental Factors

Moisture: Atmosphere humidity and precipitation in the form of rain, fog and dew affect seasonal and local incidence of diseases. Soil moisture determines the prevalence of infection and severity of diseases due to soil borne pathogens. Germination and initial development of pathogens need moisture for their development and severity. Downy mildew fungi need more moisture then powdery mildews for disease initiation and further development. Moisture makes the plant cells turgid which facilitates the pathogen to grow in the host.

Temperature: Temperature is the determining factor in the seasonal and regional incidence of disease and determines the geographic distribution. Late blight of potato is the disease of cooler places at higher altitude but in winter in subtropics epidemics are common. The temperature has predisposing effect on the host. Greatest damage to tobacco is caused by *Thielaviopsis basicola* at 17 to 23 °C and very little at 28 to 30 °C which is optimum for growth of tobacco. This shows the host is weakened at lower temperature. The virus diseases are also influenced by temperature. Tobacco mosaic virus produces small , necrotic, discrete lesion on tobacco plant at low temperature, the lesions increase in size with increasing temperature and at 28-35 °C they tend to coalesce.

Light: The development of some diseases is influenced by light intensity and day length affect survival of inoculums , pre-penetration process, length of inoculation period, sporulation and pathogenic effect. Cereal seedling surface due to root rot and basal stem rots at low light intensity. The expression of symptoms and severity of some virus diseases gets affected by light. Light may predispose plants to infections.

Soil pH: The alkalinity and acidity of soil limits the distribution of some diseases . The influence is on the pathogen. Potato scab is more in dry alkaline soils (pH 5.2-8.0) and club root of crucifers in wet acid soils (pH 5.7-6.2). *Fusarium oxysporium* f. sp. *pisi* causing wilt of pea is favoured by alkaline soil whereas; *F. oxysporum* f. sp. *lycopersici* causing wilt of tomato is favoured by acidic soils.

Oxygen and carbon dioxide in soil: The aeration of soil is dependent to a considerable extent on soil type, texture, soil moisture and soil cover. Oxygen has effect on diseases as it influences soil microflora. Due to lack of oxygen rapid root development and periderm formation are inhibited and are invaded by pathogens. In wet, poorly aerated soils the hosts gets predisposed to attack by *Pythium*.

Nutrients: Nutrients are essential for good and fast growth of plants and the pathogens depend upon host plants for their nutrition. Some pathogens attack low growing or weak hosts, others thrive well on fast growing hosts. N, P, K, Ca, B, Cu, Fe, Mg, Mn, Mo, S and Zn are quantity by various crops. N is needed for vigorous growers of plants. N deficient plants are attacked by pathogens at early stage by many pathogens. Excess N tends to lengthen the long vegetation period and delays maturity period susceptibility to many pathogens is increased. P promotes good root development and maturity of seed. Ca has direst effect of soil borne pathogens. Ca deficiency increases severity of root rot of pea caused by *Rhizoctonia solani.* Low level of K favours Fusarium wilt of cotton and tomato. Young rapidly growing plant tissues are more susceptible to viruses than older ones

Activities of Human: Many activities of human have direct or indirect effect on plant disease epidemic such as selection of sites; propagative materials and various disease management practices including cultural practices have a direct or indirect effect on disease epidemics. Low-lying and poorly drained field, especially if near other infected fields tend to favour the appearance and development of epidemic. The use of infected seeds, nursery stocks and other propagative materials, increase the amount of primary inoculums within the crop and thus, greatly favour the development of epidemics. Continuous monoculture, large acreage of the same variety of the crop, high level of nitrogen fertilizers, dense planting and poor sanitation all increase the possibility and severity of epidemic.

Patterns of plant disease epidemiology: Plant disease epidemics can exhibit various patterns, depending on the nature of the pathogen, host plant, environment, and time. Some common patterns observed in plant disease epidemiology are:

Monocyclic Epidemics

1. These involve a single infection cycle per season.
2. The disease progresses slowly and steadily over time.
3. Typical examples include soil-borne diseases like Fusarium wilt of flax.
4. e.g. Fusarium wilt of flax,Bean, cotton

Polycyclic Epidemics

1. These involve multiple infection cycles within a season.
2. The disease progresses rapidly, with exponential growth in disease incidence.
3. Commonly observed in airborne diseases like powdery mildew.
4. e.g., rust of wheat and Powdery mildews

Bimodal Polycyclic Epidemics

1. These exhibit two distinct peaks of disease incidence within a season.
2. Caused by pathogens that infect different plant parts at different times.

3. An example is brown rot of stone fruits, where blossoms and fruits are infected separately. This fungal disease infects blossoms early in the season and then later infects the developing fruits, resulting in two distinct peaks of disease incidence.

Spatial Patterns

1. Disease spread can exhibit different spatial patterns.
2. Some diseases spread uniformly across a field, while others form clusters or foci of infection.
3. Spatial patterns are influenced by factors like pathogen dispersal mechanisms and plant distribution.

Uniform spread: Some diseases, like certain viral infections transmitted by aphids, spread evenly throughout a field due to the uniform distribution of the vector.

Clustered spread: Soil-borne diseases like Phytophthora root rot often occur in clusters where soil conditions favour the pathogen.

Temporal Patterns

1. Disease incidence can fluctuate over time due to changes in environmental conditions, host susceptibility, or pathogen virulence.
2. Some diseases exhibit seasonal patterns, while others may be sporadic or persistent.

Seasonal: Apple scab, a fungal disease, is more prevalent during cool, wet spring weather.

Sporadic: Bacterial leaf spot of pepper can occur sporadically throughout the season depending on rainfall and temperature.

Disease Progress Curves

1. These graphs plot disease incidence or severity over time.
2. Different disease progress curves can be observed, such as linear, exponential, or logistic curves.
3. The shape of the curve provides insights into the dynamics of the epidemic.

Disease Gradients

1. These describe the change in disease incidence or severity over distance from the initial source of infection.
2. Disease gradients can be steep or gradual, depending on the dispersal mechanism of the pathogen

- **Steep gradient:** A wind-borne fungal disease like wheat stem rust may exhibit a steep gradient, with high disease incidence near the source and rapidly declining levels further away.
- **Gradual gradient:** Soil-borne diseases like Verticillium wilt of cotton often show a gradual gradient, with disease severity decreasing slowly over distance from the initial infection site.

Understanding these patterns is crucial for developing effective disease management strategies. By identifying the pattern of an epidemic, plant pathologists can predict its course, assess its potential impact, and implement appropriate control measures to minimize crop losses.

Models in Plant disease epidemiology: Mathematical models in plant disease epidemiology are tools used to describe and predict the spread of plant diseases within a population. They utilize mathematical equations and statistical methods to quantify various factors influencing disease development and progression. These models can provide valuable insights for understanding disease dynamics, predicting outbreaks, and developing effective management strategies.

There are several types of mathematical models used in plant disease epidemiology, each with its own strengths and limitations:

Empirical Models

1. Based on observed patterns and correlations between disease variables and environmental factors.
2. Simple to develop and easy to interpret.
3. Limited in their ability to predict disease spread under varying conditions.
4. Examples: Disease progress curves, logistic models.

Mechanistic Models

1. Based on biological and physical processes involved in disease development and spread.
2. More complex and require detailed knowledge of the pathosystem.
3. Provide a deeper understanding of the underlying mechanisms.
4. Examples: SIR (Susceptible-Infected-Removed. models, models incorporating host resistance and pathogen virulence.

Spatial Models

1. Account for the spatial distribution of plants and the movement of pathogens within a field or landscape.
2. Can be used to predict the spread of disease from an initial point of infection.
3. Examples: Cellular automata models, diffusion models.

Simulation Models

1. Combine various model components to create a virtual representation of the epidemic process.
2. Allow for the exploration of different scenarios and management strategies.
3. Examples: Agent-based models, Monte Carlo simulations.

The choice of model depends on the specific research question, available data, and desired level of complexity. Some common applications of mathematical models in plant disease epidemiology include:

- **Disease forecasting:** Predicting the timing and severity of disease outbreaks to guide management decisions.
- **Evaluating control strategies:** Assessing the effectiveness of different control measures like fungicides, cultural practices, or resistant cultivars.
- **Risk assessment:** Estimating the probability of disease introduction or spread to new areas.

- **Understanding disease dynamics:** Exploring the factors that influence the rate and pattern of disease spread.

By incorporating mathematical models into plant disease epidemiology, researchers and practitioners can gain valuable insights into the complex interactions between plants, pathogens, and the environment. This knowledge can ultimately lead to the development of more effective and sustainable disease management strategies.

Plant disease spread through primary and secondary inoculums

Plant disease spread involves a combination of primary and secondary inoculum, which are the sources of the pathogen that cause infection.

Primary Inoculum

- This is the initial source of the pathogen that starts a disease epidemic in a plant population.
- It can survive in the soil, plant debris, seeds, or alternative hosts during periods unfavorable for disease development.
- Primary inoculum can be introduced to a new area through various means, including wind, water, insects, or human activities.
- Once introduced, it infects susceptible plants, establishing the first infection cycle.

Examples of Primary Inoculum Sources

- **Overwintering structures:** Fungal spores or sclerotia that survive through the winter.
- **Infected seeds:** Seeds carrying the pathogen internally or externally.
- **Volunteer plants:** Plants from previous crops that harbor the pathogen.
- **Alternative hosts:** Weeds or other plants that can serve as reservoirs for the pathogen.

Secondary Inoculum

- This is the inoculum produced on the infected plants during the current growing season.
- It is responsible for the subsequent spread of the disease within the field or to neighboring fields.
- Secondary inoculum is often produced in large quantities and dispersed rapidly, leading to a rapid increase in disease incidence.

Examples of Secondary Inoculum Production

- **Spores:** Fungal or oomycete spores released from infected plant tissues.
- **Bacteria:** Bacterial cells oozing from lesions or plant surfaces.
- **Viruses:** Viral particles transmitted by vectors like insects or nematodes.

Disease Cycle and Inoculum:

The disease cycle involves a series of events:

1. Primary inoculum infects a susceptible plant.
2. The pathogen colonizes the plant tissues, causing disease symptoms.

3. The pathogen produces secondary inoculum on the infected plant.
4. Secondary inoculum is dispersed to healthy plants, starting new infection cycles.
5. The cycle repeats, leading to an epidemic if conditions are favorable.

Disease Management Strategies

Understanding the roles of primary and secondary inoculum is crucial for developing effective disease management strategies:

- **Managing Primary Inoculum**
  - Removing crop residues and volunteer plants.
  - Using disease-free seeds.
  - Controlling alternative hosts.
  - Applying seed treatments or soil fumigants.
- **Managing Secondary Inoculum**
  - Using fungicides or bactericides to protect healthy plants.
  - Removing infected plants to reduce inoculum sources.
  - Planting resistant cultivars.
  - Modifying cultural practices to create less favorable conditions for disease spread.

By targeting both primary and secondary inoculum, integrated disease management programs can effectively reduce disease incidence and minimize crop losses.

New Tools in epidemiology: The study of plant disease epidemiology has been facilitated greatly by new methods and new equipment.

Molecular tools: The most important tools of these are development and use of genetic (DNA. probe that allow the definite detection and identification of a plant pathogen within or on the surface of a plant tissue that include following tools. Polymerase chain reaction (PCR), Random amplified polymorphic DNA (RAPD..

Geographic Information system (GIS): GIS is a computer system that can be installed on any model desktop computer and is capable of assembling, storing, manipulating and displaying data referenced by Geographic coordinates. GIS is adaptable to operation of any size and data can be used at any scale from single field to an agriculture region.

Global positioning system (GPS): It consists of a hand held device that coordinate with a global system of manmade satellites and depending on the accuracy and co-ordination, provide quite accurate reading of the coordinate of the position of the device.GPS enables a specific area of the field that are affected by the pathogen, which can be visited and examined again periodically for incremental advance of the symptom.

Geostatistics: Geostatistics consists of various "geostatistical" techniques that are applied in plant disease epidemiology to characterize quantitatively spited patterns at disease development or the development of pathogen populations in space and time. These techniques have the capability to take into account the characteristics of spastically distributed variables whether they are random or systematic. V Remote sensing: Remote sensing usually refers to the use of instruments for measuring electromagnetic radiation reflected or emitted from an object. The various remote sensing instruments store data obtained from field situations and data are then printed out are analyzed.

Image Analysis: Image analysis refers to photography and electronic image analysis usually of large areas of fields or of mountains. The image or photographs are taken through aerial photography, ground based sensor data and satellite- borne and air borne sensor. Air borne multispectral scanning is studied and used widely for the surveillance of plant disease, pests and environmental stresses in agriculture.

Information Technology: This technology involves primarily the use of computers alone or in combination with other electronic devices. They help to collect data on plant diseases at various levels and various locations in a continuous manner electronic information technology can above all, describe and display spatial pattern of characteristics of different pathogens

1. What is plant disease epidemiology?

Plant disease epidemiology or epiphytology is the study of how plant diseases spread and impact plant populations. It involves analyzing the factors that influence disease transmission, development, and management. It involves changes in disease intensity over time and space in a host population.It deals with host-pathogen interactions that are dependent on abiotic and biotic environments causing disease and crop losses under man's influence .

2. Who is the father of plant epidemiology

James Edward Vanderplank is known as "Father of Plant Epidemiology", who gave the concept of horizontal and vertical resistance in plants in 1968. He coined the term "Vertifolia effect"-loss of horizontal resistance during breeding for vertical resistance in Vertifolia (German potato variety infected) disease with late blight. His particular publications were- **Plant Disease Epidemics and Control (1963); Disease Resistance in Plants (1968); Principles of Plant Infection (1975); Genetics and Molecular Basis of Plant Pathogen (1978); Host Pathogen Interaction in Plant Disease (1982).**

3. What are the main components of the disease triangle in plant epidemiology?

The disease triangle consists of three main components a susceptible host, a virulent pathogen, and disease favourable environmental conditions. Disease occurs when all three components intersect over time.

4. Who is credited with pioneering the concept of the disease triangle in plant pathology?

The concept of the disease triangle, which emphasizes the interaction of a susceptible host, a virulent pathogen, and conducive environmental conditions, was first introduced by Paul Sorauer in 1914 his book, "Plant Diseases Their Causes and Social Significance." This concept has since become a fundamental principle in plant disease epidemiology.

5. What is the difference between primary and secondary plant disease cycles?

The primary disease cycle involves direct infection of the host by the pathogen, while the secondary disease cycle typically involves the production and spread of pathogen propagules, such as spores, to initiate new infections.

6. How does the concept of disease progress relate to plant epidemiology?

Disease progress refers to the development of a disease epidemic over time. Plant epidemiologists study the rate and pattern of disease progress to understand how quickly and extensively a disease spreads within a plant population.

7. AUDPC stands for what?

AUDPC stands for "Area under the Disease Progress Curve." It is a measurement used in plant pathology and epidemiology to assess the cumulative disease severity or progress over time, often plotted on a graph to understand how a disease develops and spreads in a crop or plant population.

8. What role do vectors play in plant disease epidemiology?

Vectors are organisms, often insects or other pests that transmit plant pathogens from one host to another. They play a crucial role in the spread of many plant diseases.

9. Why is it important to measure plant disease intensity in agriculture?

Measuring plant disease intensity is crucial for assessing the extent of damage, making informed management decisions, and evaluating the effectiveness of control measures. Disease intensity is often measured as a percentage, representing the proportion of infected plant tissue relative to the total tissue examined.

10. How can the mutation rate of infectious agents affect epidemic development?

The mutation rate of infectious agents can impact epidemic development by potentially giving rise to new strains or variants that may have different characteristics, including increased virulence or altered transmission dynamics. These variants can influence the course and severity of an epidemic.

11. What are some common methods for managing plant diseases in agriculture based on epidemiological principles?

Common disease management strategies include crop rotation, planting resistant varieties, skip disease development by changing planting time, applying fungicides or pesticides, and modifying cultural practices to reduce disease pressure.

12. How can early disease detection and surveillance help in plant disease management?

Early detection allows for timely intervention and control measures, reducing the potential for disease spread and economic losses.

13. What are the ethical considerations in plant disease epidemiology research, particularly in the context of genetically modified crops and disease-resistant strains?

Ethical considerations may include issues related to biodiversity, ecological balance, and the potential for unintended consequences when implementing disease-resistant plant varieties.

14. What historic event in the mid-20th century had a profound impact on the development of plant disease epidemiology?

The discovery of the structure of DNA in the 1950s by James Watson and Francis Crick marked a significant milestone. It allowed researchers to better understand the genetic basis of plant diseases, leading to advancements in the study of plant disease resistance and epidemiology.

15. What historic plant disease outbreaks had a profound impact on agriculture and the development of plant disease management strategies?

The Irish Potato Famine (1845-1852) caused by the potato late blight pathogen (*Phytophthora infestans*) and the Southern Corn Leaf Blight epidemic in the United States (1970) significantly influenced the development of plant disease management strategies. These outbreaks highlighted the importance of disease-resistant crop varieties and diversified agricultural practices.

16. What are the favourable conditions for late blight development?

Late blight develops when the pathogen *Phytophthora infestans*, infects potato plants. Favourable conditions for late blight development include cool and wet weather, high humidity (90%), and temperatures between 10-20°C (50-68°F).

17. What environmental and climatic factors contribute to the prevalence of apple scab in Himachal Pradesh?

Apple scab is a fungal disease caused by the pathogen *Venturia inaequalis*. Apple scab incidence in Himachal Pradesh is influenced by several environmental factors, including humidity, rainfall, and temperature. Rain and high humidity create favourable conditions for the development and spread of the scab pathogen. Understanding these factors is crucial for predicting disease outbreaks and implementing preventive measures. EPIVEN, a computer based programme is used to disease forecasting.

18. What are some disease management strategies employed in Himachal Pradesh to control apple scab?

Disease management strategies for apple scab in Himachal Pradesh include the use of fungicides, sanitation practices (removing infected leaves and fruit), application of 5% urea at end of harvesting for quick destruction of the infected leaves in dormant season and the planting of scab-resistant apple varieties. Additionally, monitoring and early detection through disease assessment play a vital role in guiding timely interventions to minimize scab-related losses in apple orchards.

19. What is citrus canker, and how does it impact the epidemiology of citrus diseases?

Citrus canker, caused by the bacterium *Xanthomonas axonopodis*, is a contagious disease affecting citrus trees. It is important epidemiologically because it can lead to the quarantine and destruction of infected trees to prevent its spread, impacting citrus production and trade.

20. Why is black sigatoka a significant epidemiological concern in banana cultivation?

Black sigatoka, caused by the fungus *Mycosphaerella fijiensis*, is a major threat to banana crops. It is important epidemiologically because it can lead to significant yield losses and requires intensive fungicide applications, impacting both the environment and the cost of banana production.

21. What is the role of host resistance in managing epidemiologically important plant diseases?

Host resistance, either through natural genetic resistance or the development of resistant crop varieties, plays a crucial role in managing many epidemiologically important plant diseases. It reduces the susceptibility of host plants to infection and can slow down the spread of diseases, contributing to disease control and management efforts.

22. Why is the epidemiological spread of rice blast disease a major concern for rice cultivation in India?

Rice blast disease, caused by the fungus *Magnaporthe oryzae*, is a significant threat to rice production in India. It can lead to substantial yield losses and reduce the quality of rice grains. The disease is particularly concerning due to its ability to rapidly spread during monsoon seasons and affect large rice-growing regions.

23. How does the spread of coconut lethal yellowing disease impact the coconut industry in India?

Coconut lethal yellowing disease, caused by phytoplasmas, affects coconut palms in several Indian states. It poses a major threat to the coconut industry as it leads to the death of coconut palms, impacting coconut production and livelihoods in coastal regions.

24. Who is known for pioneering the use of mathematical models in plant disease epidemiology?

Sir Ronald A. Fisher, a British statistician and geneticist, made significant contributions to the use of mathematical models in understanding the dynamics of plant disease epidemics. His work laid the foundation for modern epidemiological modeling in plant pathology. Mathematical models help predict disease outbreaks, assess the impact of control measures, and understand the dynamics of disease spread in plant populations.

25. What historic events or developments led to the emergence of molecular techniques in plant disease epidemiology?

The advent of polymerase chain reaction (PCR) technology in the 1980s by Kary Mullis, revolutionized plant disease epidemiology by enabling researchers to detect and study plant pathogens at the molecular level. This breakthrough allowed for more accurate and rapid identification of pathogens.

26. What are plant epidemiological models, and how do they work in the context of Indian agriculture?

Plant epidemiological models are mathematical tools used to simulate the spread and impact of plant diseases. In India, these models help researchers and policymakers understand how diseases spread in various crops, predict disease outbreaks, and design effective control strategies. Climate and weather data are crucial inputs for plant epidemiological models in India. They help model the effect of temperature, humidity, and rainfall on disease development and transmission, allowing for more accurate predictions and control strategies.

27. What is the "Epidemic Triangle" in plant disease epidemiology?

The Epidemic Triangle model represents the interaction of three components a susceptible host, a virulent pathogen, and a conducive environment, all of which are necessary for a disease outbreak to occur.

28. What is the difference between an endemic and an epidemic disease pattern?

Endemic diseases are consistently present at low levels in a region, while epidemics represent a sudden and significant increase in disease cases, often exceeding what is considered typical.

29. What is the relationship between disease severity and yield losses?

Disease severity is directly related to yield losses; as disease severity increases, crop yields tend to decrease.

30. What is the "Threshold Model" in plant disease epidemiology?

The Threshold Model predicts disease incidence based on the relationship between the number of pathogens present and the host's susceptibility. When pathogen numbers exceed a certain threshold, an epidemic occurs.

31. What is "Quarantine" in plant disease management?

Quarantine is a legal and regulatory practice aimed at restricting the movement of potentially infected plant material or hosts to prevent the introduction or spread of plant diseases in new regions.

32. When was remote sensing was used first time in India and on which crop?

The first experiment that used remote sensing was on coconut root wilt disease by 'Prof. Rama Pisharoty' in Kerela in early 1970s. He was regarded as "Father of remote sensing in India", who used Hasselblad camera in seven different combinations of wave bands, flown by using helicopter.

33. What role do remote sensing and GIS (Geographic Information Systems) play in plant epidemiological modeling in India?

Remote sensing and GIS technologies are used to collect and analyze spatial data related to crop health and disease distribution. These tools help monitor the spread of diseases across different regions and guide targeted interventions, such as pesticide applications or quarantine measures.

34 What are some examples of plant diseases in India where epidemiological models have been used effectively?

Epidemiological models have been applied to diseases like rice blast, wheat rust, tomato leaf curl virus, and potato late blight in India. These models have helped researchers understand disease patterns and develop strategies to mitigate their impact.

35 What are different computer simulation models used against late blight of potato for plant disease epidemiology?

BLIGHT CAST (Karuse *et al.*, 1975), SIMCAST (Grunwald, 2002), NEGFRY (Runno and Kopple, 2002) and POTATOES are different computer based programmes used against late blight of potato.

36. How TOMEX works as an expert systems in plant disease epidemiology?

It uses 78 questions and 87 photographs already setup in the system for diagnosis of 37 diseases of tomato.

37. What is critical point model? How it is used in plant disease epidemiology?

The Critical Point Model is a concept used to describe the relationship between the severity of a disease and the potential for an epidemic. In this model, there is a "critical point" or threshold, below which an epidemic is unlikely, and above which the epidemic is more likely to occur. The critical point is determined by factors such as the pathogen's virulence, the host plant's susceptibility, environmental conditions, and the availability of inoculums, ex. Rice blast, wheat stem rust, and potato cyst nematode.

38. What are some common mechanisms of dispersal for plant pathogens?

Common mechanisms of dispersal for plant pathogens include wind, rain splash, insects, birds, animals, contaminated tools, and human activities.

39. How does rain splash contribute to the dispersal of plant pathogens and name these diseases caused by rain dispersal?

Raindrops can splash pathogens from infected plant surfaces onto nearby healthy plants, potentially leading to new infections. Coffee rust (*Hemileia vastatrix*), bacterial blight (*Xanthomonas compestris* pv. *oryzae*), and streak disease of paddy (*Xanthomonas oryzicola*. are water splash dispersed diseases, while *Plasmodiophora brassicae*, *Phytophthora infestansans*, *Colletotrichum falcatum* and *Fusarium* spp. are the water flow dispersed pathogens. Water dispersal of spores is called hydrochory.

40. What role do human activities play in the dispersal of plant pathogens?

Human activities, such as the movement of infected plant materials, equipment, or soil, can inadvertently spread plant pathogens to new areas and plant populations.

41. Name those plant diseases which can dispersed by farm and wild animals?

Citrus canker, fire blight, Pierce's disease, seed borne diseases, crown gall, powdery mildew, rust, wart disease and flag smut of wheat are some diseases which can be dispersed by insects, man, nematodes and wild animals.

42. What is the "Plant Variety Protection Act" (PVPA.?

The PVPA is a U.S. law that grants intellectual property rights to plant breeders, allowing them to protect and license their new plant varieties. It encourages the development of disease-resistant plant varieties.

43. What is the "Plant Protection Act" (PPA.?

The PPA is a U.S. law that provides the authority to regulate the import, export, and interstate movement of plants, plant products, and organisms that can harm agriculture, including plant pathogens. It helps prevent the spread of plant diseases.

44. What is the relationship between Soil pH and nutrient availability to plants?

Soil pH can affect the solubility of essential nutrients. For example, acidic soils can limit the availability of nutrients like calcium, which is important for plant cell wall strength and disease resistance.

45. How can soil pH adjustments be used in disease management?

Soil pH adjustments through liming (raising pH) or acidification (lowering pH) can be employed to create conditions less favourable for specific plant pathogens, helping reduce disease incidence.

46. How can soil nutrient imbalances contribute to disease outbreaks?

Proper nutrient management can enhance a plant's immune system, making it more resistant to diseases. Excessive levels of certain nutrients, like nitrogen, can promote lush, succulent plant growth, which is more susceptible to diseases. Nutrient imbalances can disrupt the plant's ability to defend against pathogens.

47. How does monitoring primary and secondary inoculum sources aid in disease forecasting?

Regular monitoring of primary and secondary inoculum sources helps forecasters assess disease risk and make informed recommendations for disease management strategies.

48. How can spore trap data aid in disease forecasting?

A spore trap is a device designed to capture and collect airborne spores released by plant pathogens, providing valuable data for disease monitoring and management. It uses air suction to capture spores from the atmosphere typically on a microscope slide or a sticky tape. Analysis of spore trap data can provide insights into the timing and severity of disease epidemics, helping growers make informed decisions on disease management strategies.

49. What is the importance of understanding the epidemic pattern for disease management?

Understanding the pattern helps in timing disease control measures effectively. For example, early intervention during the initial outbreak or exponential phase can be more successful in limiting the epidemic.

50. What factors can influence the accuracy of yield loss assessments due to plant diseases?

Factors such as timing of assessment, disease resistance in the crop, and environmental conditions can affect the accuracy of yield loss estimates.

Quick Revision with Short Question Answers

1. What is plant disease epidemiology?

Plant disease epidemiology is the study of disease development in plant populations, focusing on the factors influencing disease outbreaks and spread.

2. What is an epidemic?

An epidemic is a widespread occurrence of a disease in a plant population over a large area within a short period.

3. How are epidemics different from pandemics?

Epidemics are localized to a specific region or country, while pandemics spread across continents or subcontinents.

4. What are the main objectives of studying plant disease epidemiology?

The main objectives include understanding the causes of diseases, studying plant-pathogen interactions, comprehending disease development mechanisms, and devising effective disease management strategies.

5. What are the primary causes of plant diseases?

Plant diseases can result from direct cell damage, toxins affecting metabolism, nutrient and water competition, and interference with nutrient uptake.

6. What is the disease triangle?

The disease triangle illustrates the three main components necessary for disease development a susceptible host, a virulent pathogen, and a favorable environment.

7. How does the disease pyramid expand on the disease triangle concept?

The disease pyramid adds two more factors to the disease triangle time and human activities, acknowledging their influence on disease development.

8. What are monocyclic diseases?

Monocyclic diseases are characterized by pathogens completing only one disease cycle per year.

9. How do monocyclic diseases progress over time?

Monocyclic diseases typically exhibit a saturation curve pattern, where disease incidence increases and eventually plateaus as the season progresses.

10. Can you provide examples of monocyclic diseases?

Examples include blackleg of potato, Verticillium wilt, and cereal cyst nematode.

11. What are polycyclic diseases?

Polycyclic diseases involve pathogens that undergo multiple disease cycles within a single growing season.

12. How do polycyclic diseases spread?

Polycyclic diseases often spread through the production of large quantities of asexual spores, leading to secondary infections and rapid disease multiplication.

13. What is the typical disease progress curve for polycyclic diseases?

Polycyclic diseases usually follow a sigmoidal curve, with an initial slow increase in disease incidence, followed by a rapid exponential phase and eventual stabilization.

14. Can you name some examples of polycyclic diseases?

Examples include downy mildews, powdery mildews, late blight of potato, and various leaf spots and blights.

15. How do polyetic diseases differ from monocyclic and polycyclic diseases?

Polyetic diseases have multi-year cycles, where pathogens may take several years to produce inoculum and initiate new infections.

16. What are some examples of polyetic diseases?

Examples include Dutch elm disease, citrus tristeza, and certain fungal vascular wilts.

17. What is the apparent infection rate?

The apparent infection rate represents the rate at which new infections appear in a plant population over time.

18. How is the apparent infection rate influenced by the type of disease cycle?

In monocyclic diseases, the apparent infection rate is relatively constant, while in polycyclic diseases, it increases exponentially due to the repeated cycles of infection and spore production.

19. How can knowledge of disease cycles be used for disease management?
Understanding disease cycles helps in developing targeted management strategies, such as reducing initial inoculum for monocyclic diseases and focusing on reducing the infection rate for polycyclic diseases.

20. What are some host factors affecting disease development?

Host factors include the level of genetic resistance or susceptibility, genetic uniformity, crop type, age of the plant, and the introduction of new host plants.

21. How do pathogen factors influence epidemics?

Pathogen factors include the level of virulence and aggressiveness, the quantity of inoculum near the host, the type of reproduction, the ecology of the pathogen, and its mode of spread.

22. What environmental factors contribute to disease epidemics?

Environmental factors like temperature, humidity, rainfall, and the presence of vectors can significantly impact disease development.

23. How can human activities affect plant disease epidemics?

Human activities such as site selection, choice of propagating materials, cultural practices, and disease control measures can either promote or hinder disease outbreaks.

24. What is a disease progress curve?

A disease progress curve graphically represents the development of a disease over time, showing the increase in disease incidence or severity.

25. What is a disease gradient curve?

A disease gradient curve illustrates the spatial pattern of disease spread, showing how disease intensity varies across distances from the initial infection point.

26.What is virulence?

Virulence refers to the degree of pathogenicity or the capacity of a pathogen to cause disease on a host plant.

27. How does the quantity of inoculum affect disease development?

A higher quantity of initial inoculum near a susceptible host plant generally leads to a more severe and rapid disease outbreak.

28. What are the different types of reproduction in pathogens, and how do they influence epidemics?

Pathogens can reproduce sexually or asexually. Asexual reproduction, often through spores, allows for rapid multiplication and spread of the pathogen, leading to more severe epidemics.

29. How does the ecology of a pathogen contribute to epidemics?

The ecology of a pathogen, including its survival mechanisms, preferred hosts, and dispersal methods, plays a crucial role in determining its ability to cause epidemics.

30. What are the primary modes of pathogen spread?

Pathogens can spread through various means, including wind, water, soil, seed, insects, human activities, and agricultural machinery.

31. How does temperature influence disease development?

Temperature affects both host resistance and pathogen growth. High temperatures can weaken host defenses and favor pathogen development, while low temperatures can slow down pathogen growth and reduce disease spread.

32. What role does humidity play in plant disease epidemics?

Humidity is essential for the germination of spores and the penetration of pathogens into host plants. High humidity often creates a favorable environment for disease development.

33. How does rainfall contribute to disease spread?

Rainfall can disperse pathogen inoculum, create a conducive environment for infection, and facilitate the movement of vectors carrying pathogens.

34. What is the significance of leaf wetness in disease development?

Leaf wetness is a critical factor for many foliar diseases, as it provides the necessary moisture for spore germination and infection.

35. How does foliage density impact disease epidemics?

Dense foliage can create a microclimate with high humidity, promoting disease development and spread.

36. What is the difference between host resistance and susceptibility?

Host resistance refers to the plant's ability to defend itself against pathogen infection, while susceptibility indicates a lack of effective defense mechanisms.

37. How does genetic uniformity in host plants affect epidemics?

Genetically uniform plant populations, such as those resulting from vegetative propagation or monoculture, are more vulnerable to widespread disease outbreaks due to the lack of genetic diversity and resistance.

38. What role does crop type play in disease epidemics?

Different crop types exhibit varying levels of susceptibility to specific pathogens. Annual crops and those with foliar or fruit diseases tend to experience more rapid disease development compared to perennial woody crops.

39. How does the age of a host plant influence its susceptibility to disease?

Plant susceptibility to diseases can change with age. Some plants are more susceptible during their growth stages and become more resistant as they mature.

40. What is the impact of introducing new host plants on disease epidemics?

Introducing new host plants can disrupt existing disease dynamics and potentially lead to the emergence of new diseases or the spread of existing pathogens to previously unaffected areas.

41. How do human activities influence the time factor in disease development?

Human activities such as planting date, crop rotation, and irrigation practices can directly impact the timing of disease occurrence and the duration of epidemics. For example, early planting may expose crops to higher disease pressure during their susceptible growth stages.

42. What is the role of time in the disease pyramid?

Time is a critical component of the disease pyramid as it represents the duration of the epidemic and the time it takes for disease to develop and spread. The time factor interacts with host, pathogen, and environmental factors to determine the overall disease dynamics.

43. How do seasonal variations affect disease development?

Seasonal variations in temperature, humidity, and rainfall patterns create conducive conditions for different diseases at specific times of the year. Understanding these seasonal patterns helps in predicting disease outbreaks and implementing timely management strategies.

44. What is the significance of the appearance of vectors in disease epidemics?

The presence and activity of insect vectors, which transmit pathogens, are often influenced by seasonal changes. The timing of vector emergence and their population dynamics can significantly impact disease spread.

45. How can human interventions disrupt the natural progression of disease epidemics?

Human interventions, such as the application of fungicides or the introduction of resistant crop varieties, can alter the course of disease epidemics by reducing pathogen populations or slowing down disease spread.

46. How does the management of monocyclic diseases differ from that of polycyclic diseases?

Management of monocyclic diseases primarily focuses on reducing the initial inoculum, as there is only one infection cycle per season. In contrast, polycyclic diseases require strategies to minimize both the initial inoculum and the rate of disease spread throughout the season.

47. What are some effective management practices for monocyclic diseases?

Effective practices for monocyclic diseases include using disease-free seeds or planting material, crop rotation, sanitation to remove infected plant debris, and appropriate fungicide applications to target the primary inoculum.

48. How can polycyclic diseases be managed effectively?

Managing polycyclic diseases involves an integrated approach, including the use of resistant varieties, cultural practices to reduce disease spread (e.g., proper plant spacing, pruning), and timely fungicide applications to protect plants during critical growth stages.

49. What is the role of host resistance in managing both monocyclic and polycyclic diseases?

Host resistance is a crucial factor in disease management for both types of diseases. Resistant varieties can significantly reduce disease incidence and severity, limiting the need for other control measures.

50. How does the timing of management practices impact disease control?

The timing of management practices is critical for effective disease control. For monocyclic diseases, early-season interventions are essential to reduce the primary inoculum. In polycyclic diseases, timely applications of fungicides or other control measures are necessary to prevent disease establishment and subsequent spread.

51. How does the management of monocyclic diseases differ from that of polycyclic diseases?

Management of monocyclic diseases primarily focuses on reducing the initial inoculum, as there is only one infection cycle per season. In contrast, polycyclic diseases require strategies to minimize both the initial inoculum and the rate of disease spread throughout the season.

52. What are some effective management practices for monocyclic diseases?

Effective practices for monocyclic diseases include using disease-free seeds or planting material, crop rotation, sanitation to remove infected plant debris, and appropriate fungicide applications to target the primary inoculum.

53. How can polycyclic diseases be managed effectively?

Managing polycyclic diseases involves an integrated approach, including the use of resistant varieties, cultural practices to reduce disease spread (e.g., proper plant spacing, pruning), and timely fungicide applications to protect plants during critical growth stages.

54. What is the role of host resistance in managing both monocyclic and polycyclic diseases?

Host resistance is a crucial factor in disease management for both types of diseases. Resistant varieties can significantly reduce disease incidence and severity, limiting the need for other control measures.

55. What is a disease progress curve?

A disease progress curve is a graphical representation of how a plant disease develops within a population over time. It shows the increase in disease incidence or severity as the pathogen spreads and infects more plants.

56. What is the typical shape of a disease progress curve for monocyclic diseases?

For monocyclic diseases, the disease progress curve typically follows a saturation curve pattern. This means that the disease incidence increases rapidly at first and then gradually levels off as the season progresses. The leveling off occurs because the pathogen has a limited amount of initial inoculum and can only complete one disease cycle.

57. How does the disease progress curve of polycyclic diseases differ from that of monocyclic diseases?

The disease progress curve of polycyclic diseases usually exhibits a sigmoidal (S-shaped. curve. It starts with a slow increase in disease incidence, followed by a rapid exponential phase where the disease spreads quickly due to multiple disease cycles and the production of secondary inoculum. Eventually, the curve stabilizes as the disease reaches its maximum potential or environmental conditions become less favorable.

58. What factors influence the shape of disease progress curves?

Several factors can influence the shape of disease progress curves, including the type of disease cycle (monocyclic or polycyclic., the virulence of the pathogen, the susceptibility of the host plant, environmental conditions, and the effectiveness of disease management practices.

59. How can disease progress curves be used in plant disease epidemiology?

Disease progress curves provides valuable information for understanding the dynamics of disease spread, predicting disease outbreaks, and evaluating the effectiveness of disease control strategies. By analyzing the shape of the curve, epidemiologists can gain insights into the factors driving disease development and make informed decisions about disease management interventions.

6

Fungicides, Fungicide Resistance Classification and Mode of Action

Fungicides are chemical compounds or biological organisms used to kill or inhibit the growth of fungi and their spores. They are essential tools in agriculture and horticulture for protecting crops and plants from fungal diseases, which can cause significant yield losses and economic damage.

Fungistat: Some chemicals do not kill the fungal pathogens. But they simply arrest the growth of the fungus temporarily. These chemicals are called fungistat and the phenomenon of temporarily inhibiting the fungal growth is termed as fungistatis. **Antisporulant** Some other chemicals may inhibit the spore production without affecting the growth of vegetative hyphas and are called as 'Antisporulant'. Eventhough, the antisporulant and fungistatic compounds do not kill the fungi, they are included under the broad term fungicide because by common usage, the fungicide has been defined as a chemical agent which has the ability to reduce or prevent the damage caused to plants and their products

Protectant As the name suggests, protectant fungicides are prophylactic in their behaviour. Fungicide which is effective only if applied prior to fungal infection is called a protectant, eg., Zineb, Sulphur.

Therapeutant Fungicide which is capable of eradicating a fungus after it has caused infection and there by curing the plant is called chemotherapeutant. eg. Carboxin, Oxycarboxin antibiotics like Aureofungin.

Eradicant Eradicant are those which remove pathogenic fungi from an infection court (area of the host around a propagating unit of a fungus in which infection could possibly occur). eg. Organic mercurials, lime sulphur, dodine etc. These chemicals eradicate the dormant or active pathogen from the host. They can remain effective on or in the host for some time.

Fungicides can be classified in several ways, including:

1 Based on General uses

The fungicides can also be classified based on the nature of their use in managing the diseases.

a. Seed protectants: Eg. Captan, thiram, organomercuries carbendazim, carboxin etc.
b. Soil fungicides (preplant): Eg. Bordeaux mixture, copper oxychloride, Chloropicrin, Formaldehyde Vapam, etc.,

c. Soil fungicides: Eg. Bordeaux mixture, copper oxy chloride, (for growing plants) Captan, PCNB, thiram etc.
d. Foliage and blossom: Eg. Captan, ferbam, zineb, protectants like mancozeb, chlorothalonil etc.
e. Fruit protectants: Eg. Captan, maneb, carbendazim, mancozeb etc.
f. Eradicants: Eg. Organomercurials, lime sulphur, etc.
g. Tree wound dressers: Eg. Bordeaux paste, chaubattia paste, etc.
h. Antibiotics: Eg. Actidione, Griseofulvin, Streptomycin, Streptocycline, etc.,
i. General purpose spray and dust formulations..

2. Based on Chemical Nature

- **Inorganic fungicides:** These are typically simple compounds containing copper, sulfur, or mercury. Examples include Bordeaux mixture (copper sulfate and lime) and elemental sulfur.
- **Organic fungicides:** These are more complex carbon-based compounds. Examples include dithiocarbamates, triazoles, strobilurins, and benzimidazoles.

Sulphur Fungicides

Use of sulphur in plant disease control is probably the oldest one and can be classified as inorganic sulphur and organic sulphur. Inorganic sulphur is used in the form of elemental sulphur or as lime sulphur. Elemental sulphur can be either used as dust or wettable sulphur, later being more widely used in plant disease control. Sulphur is best known for its effectiveness against powdery mildew of many plants, but also effective against certain rusts, leaf blights and fruit diseases. Sulphur fungicides emit sufficient vapour to prevent the growth of the fungal spores at a distance from the area of deposition. This is an added advantage in sulphur fungicides as compared to other fungitoxicants. Organic compounds of sulphur are now widely used in these days. All these compounds, called as 'carbamate fungicides', are derivatives of Dithiocarbamic acid, Dithiocarbamates are broadly grouped into two, based on the mechanism of action.

Dithiocarbamates

Monoalkyl Dithiocarbamates: Eg. Zineb, Maneb, Mancozeb, Nabam, Vapam Dialkyl Dithiocarbamates: Eg. Thiram, Ziram, Ferbam

Copper Fungicides

The fungicidal action of copper was mentioned as early as 1807 by Prevost against wheat bunt disease *(Tilletia caries),* but its large scale use as a fungicide started in 1885 after the discovery of Bordeaux mixture by Millardet in France. The mixture of copper sulphate and lime was effective in controlling downy mildew of grapevine caused by *Plasmopara viticola* and later, late blight of potato *(Phytophthora infestans).* Some other copper sulphate preparations later developed were Bordeaux paste, Burgundy mixture and Cheshunt compound which are all very effectively used in the control of several plant diseases.

Mercury Fungicides

Mercury fungicides can be grouped as inorganic and organic mercury compounds. Both the groups are highly fungitoxic and were extensively used as seed treatment chemicals against seed borne diseases. Inorganic compounds show bactericidal properties also. However, due to their residual toxicity in soil and plants and their extreme toxicity nature to animal and human beings, the use of mercury fungicides banned.

Heterocyclic Nitrogen Compounds

Heterocyclic nitrogen compounds are mostly used as foliage and fruits protectants. Some compounds are very effectively used as seed dressers.

3. Based on Mode of Action

- **Protectant fungicides:** These form a protective barrier on the plant surface, preventing fungal spores from germinating or penetrating the plant tissue. They are typically applied before infection occurs. Examples include mancozeb and chlorothalonil.
- **Systemic fungicides:** These are absorbed by the plant and translocated to different parts, providing protection from within. They can be used to control established infections. Examples include triazoles and strobilurins.
- **Curative fungicides:** These can stop or slow down the growth of a fungus after infection has occurred. They are often used in combination with protectant fungicides. Examples include some triazoles and strobilurins.

4. Based on Mobility within the Plant

- **Contact fungicides:** These remain on the plant surface and do not penetrate the plant tissue. They protect the plant where they are applied but do not provide systemic protection. Examples include mancozeb and chlorothalonil.
- **Trans laminar fungicides:** These penetrate the leaf surface and redistribute within the leaf, providing protection to both the upper and lower leaf surfaces. Examples include some strobilurins.
- **Systemic fungicides:** These are absorbed by the plant and translocated throughout the plant, providing protection to all parts of the plant. Examples include triazoles and some strobilurins.

5. Based on Fungicide Resistance Action Committee

- FRAC classifies fungicides based on their mode of action, assigning a unique code to each group. This classification helps in managing fungicide resistance by ensuring the rotation of fungicides with different modes of action.

Fungicide Classification Chart

Classification Basis	**Category**	**Examples**	**Description**
Chemical Nature	Inorganic	Copper sulfate, sulfur	Simple compounds, often containing metals.
	Organic	Triazoles, strobilurins	Complex carbon-based compounds.

Mode of Action	Protectant	Mancozeb, chlorothalonil	Form a protective barrier on the plant surface.
	Systemic	Triazoles, some strobilurins	Absorbed by the plant and translocated to different parts.
	Curative	Some triazoles, some strobilurins	Stop or slow down fungal growth after infection.
Mobility within Plant	Contact	Mancozeb, chlorothalonil	Remain on plant surface, no penetration.
	Translaminar	Some strobilurins	Penetrate and redistribute within the leaf.
	Systemic	Triazoles, some strobilurins	Absorbed and trans located throughout the plant.
FRAC Code	1, 2, 3, etc.	Various fungicides	Classification based on mode of action, aids in resistance management.

Fungicide resistance is the decreased sensitivity or insensitivity of a fungal pathogen to a fungicide that was once effective in controlling it. This means that the fungicide no longer works as well or doesn't work at all against the target fungus.

Fungicide Resistance

- **What it is:** Decreased effectiveness of a fungicide against a target fungus.
- **Causes**
  - Natural selection of pre-existing resistant fungi.
  - Repeated use of the same fungicide.
  - Single-site mode of action fungicides.
  - High fungal reproduction rate.
- **How to avoid:**
  - Integrated Pest Management (IPM) approach.
  - Rotate fungicides with different modes of action.
  - Tank mix or alternate fungicides.
  - Use multi-site mode of action fungicides.
  - Apply fungicides correctly.
  - Monitor for resistance.

1. What is fungicide resistance?

Fungicide resistance occurs when fungi develop the ability to survive or grow in the presence of fungicides that were previously effective.

2. What factors contribute to the development of fungicide resistance?

Factors include overuse or misuse of fungicides, genetic mutations in the fungal population, and selection pressure.

3. How can farmers reduce the risk of fungicide resistance?

Farmers can rotate different modes of action, use integrated pest management (IPM) practices, and follow recommended fungicide application rates.

4 What is cross-resistance in fungicide resistance?

Cross-resistance is when resistance to one fungicide leads to resistance to other fungicides with a similar mode of action.

5. Can fungicide resistance be managed once it occurs?

It is challenging but not impossible to manage fungicide resistance. Strategies include using alternative fungicides and cultural practices.

6. What is the mode of action of captan fungicide?

Captan is a contact fungicide that disrupts various fungal processes, including spore germination and mycelial growth.

7. Which group of fungicides includes mancozeb?

Mancozeb belongs to the ethylene bisdithiocarbamates (EBDCs) group of fungicides.

8 What is the primary use of neonicotinoid fungicides?

Neonicotinoid fungicides are primarily used as seed treatments to protect plants from soil-borne pathogens.

9 How do dicarboximide fungicides work?

Dicarboximide fungicides disrupt fungal cell membrane function by inhibiting the production of fatty acids.

10 What is the mode of action of QoI inhibitors (uinone outside inhibitors)?

QoI inhibitors block the site of the mitochondrial electron transport chain in fungi.

11 What is the primary function of fungicides in controlling fungal diseases?

A Fungicides primarily work to inhibit or kill fungal pathogens, preventing or reducing disease development in plants.

12 How do contact fungicides exert their mode of action?

Contact fungicides act upon direct contact with fungal spores and mycelium by disrupting cell membranes, inhibiting spore germination, or interfering with fungal growth.

13 Explain the mode of action of systemic fungicides.

Systemic fungicides are absorbed by the plant and translocate within the plant tissues. They inhibit fungal growth from within, providing long-lasting protection.

14 What is translaminar activity in fungicides?

Translaminar fungicides can move from the upper leaf surface (adaxial) to the lower leaf surface (abaxial) or vice versa, providing better coverage and protection.

15. How do multi-site fungicides differ from single-site fungicides in terms of mode of action?

Multi-site fungicides target multiple sites or biochemical processes within the fungal cell, reducing the likelihood of resistance development, while single-site fungicides have a specific mode of action.

16. What is the mode of action of ergosterol biosynthesis inhibitors among fungicides?

Ergosterol biosynthesis inhibitors disrupt the production of ergosterol, a crucial component of fungal cell membranes, leading to membrane dysfunction and fungal cell death.

17. Explain the mode of action of QoI inhibitors (Quinone outside inhibitors) among fungicides.

QoI inhibitors block the quinol oxidation (o) site in the mitochondrial electron transport chain of fungi, disrupting energy production and ultimately killing the fungus.

18. How do strobilurin fungicides interfere with fungal pathogens?

Strobilurin fungicides inhibit fungal respiration by targeting the cytochrome b complex in the mitochondrial electron transport chain, leading to reduced energy production and fungal death.

19. What role do succinate dehydrogenase inhibitors (SDHIs) play in the mode of action of some fungicides?

SDHIs block the function of succinate dehydrogenase, an enzyme involved in the citric acid cycle, disrupting energy production and causing fungal cell death.

20. How does the mode of action of morpholine fungicides differ from other fungicide groups?

Morpholine fungicides interfere with lipid biosynthesis by disrupting the formation of fungal cell membranes, leading to cell death.

21. What are some fungicides that target the RNA polymerase enzyme in fungi?

Fungicides like thiophanate-methyl and pyrimethanil inhibit RNA synthesis in fungi, disrupting vital cellular processes.

22. How do protectant fungicides work differently from systemic fungicides in terms of mode of action?

Protectant fungicides form a protective barrier on the plant surface, preventing fungal spores from germinating or penetrating the plant, while systemic fungicides are absorbed into the plant tissues to inhibit fungal growth from within.

23. What is the mode of action of biological fungicides, and how does it differ from chemical fungicides?

Biological fungicides contain living organisms or their derivatives, which may act by parasitizing, competing with, or producing compounds toxic to fungal pathogens. Their mode of action is biological and typically less specific than chemical fungicides.

24. How do fungicides with different modes of action collectively contribute to integrated disease management?

Using fungicides with diverse modes of action reduces the risk of resistance development in fungal populations and enhances disease control effectiveness in integrated disease management programs.

25. Explain the classification of fungicides as protectant, eradicant, or curative.

Fungicides are classified as protectant when they prevent infection, eradicant when they stop an existing infection, and curative when they reduce disease symptoms after infection has occurred.

26. What are the two main categories of fungicides based on their persistence in the environment?

Fungicides are categorized as either residual or non-residual based on their persistence in the environment. Residual fungicides remain active for an extended period, while non-residual fungicides have a shorter duration of activity.

27. How are fungicides classified based on their mobility within plants?

Fungicides are classified as xylem-mobile or phloem-mobile based on their ability to move within plant tissues through the xylem or phloem vascular systems, respectively.

28. What is the difference between inorganic and organic fungicides in terms of classification?

Inorganic fungicides are typically mineral-based compounds, while organic fungicides are carbon-based compounds derived from natural or synthetic sources.

29. How are fungicides classified based on their selectivity for specific fungi or diseases?

Fungicides can be classified as broad-spectrum, which target a wide range of fungi, or narrow-spectrum, which are specific to particular fungal pathogens or diseases.

30. Explain the classification of fungicides as single-site or multi-site.

Fungicides are categorized as single-site when they target a specific biochemical process or site within the fungal cell and multi-site when they affect multiple sites, reducing the risk of resistance development.

31. What are the main categories of biofungicides in fungicide classification?

Biofungicides are categorized into three main groups microorganisms (e.g., bacteria, fungi), botanical extracts, and biochemicals (e.g., enzymes, peptides) based on their origin and mode of action.

32. How are fungicides classified based on their application method?

Fungicides can be classified as foliar fungicides when applied to plant foliage, soil fungicides when applied to the soil, and seed treatment fungicides when applied to seeds before planting.

33. What is the classification of fungicides as protectant or systemic based on their movement within plants?

Protectant fungicides remain on the plant surface and protect against fungal infections, while systemic fungicides are absorbed and translocated within the plant, providing internal protection.

34. Explain the classification of fungicides as curative or preventive based on their application timing.

Curative fungicides are applied after infection to reduce disease symptoms, while preventive fungicides are applied before infection to prevent disease development.

35. How are fungicides classified as broad-spectrum or specific in terms of target pathogens?

Broad-spectrum fungicides are effective against a wide range of fungal pathogens, whereas specific fungicides target particular fungal species or diseases.

36. What is the classification of fungicides as protectant or eradicant based on their action against existing infections?

Protectant fungicides prevent new infections but do not eradicate existing ones, whereas eradicant fungicides can stop or eliminate ongoing infections.

37. How does fungicide resistance develop at the genetic level in fungal populations?

Fungicide resistance often arises due to genetic mutations in fungal populations that confer the ability to survive or grow in the presence of fungicides.

38. What is the difference between target-site resistance and non-target-site resistance in fungicide resistance?

Target-site resistance involves mutations in the specific molecular target of the fungicide, while non-target-site resistance involves mechanisms that reduce the fungicide's effectiveness, such as increased detoxification or reduced uptake.

39. What is the importance of using mixtures or tank-mixes of fungicides in resistance management?

Mixing fungicides with different modes of action reduces the likelihood of resistance development, as it targets multiple sites within the fungi and makes it more challenging for resistant strains to emerge.

40. How can the frequency of fungicide applications influence resistance development?

Frequent or consecutive applications of the same fungicide increase selection pressure on the fungal population, promoting the development of resistance.

41. How can the use of alternative disease control methods help manage fungicide resistance?

Alternative methods like biological control, disease-resistant crop varieties, and cultural practices can reduce the reliance on fungicides and slow the development of resistance.

42. What is the role of monitoring in managing fungicide resistance?

Regular monitoring helps detect fungicide resistance early, allowing growers to adjust their management strategies and select more effective fungicides.

43. How can genetic diversity in fungal populations contribute to fungicide resistance?

Genetic diversity increases the likelihood that some individual fungi will have mutations conferring resistance, making it more challenging to control the entire population with fungicides.

44. What are some potential consequences of fungicide resistance if left unmanaged?

If left unmanaged, fungicide resistance can result in reduced disease control, increased crop losses, higher production costs, and the need for more potent (and potentially more environmentally harmful) fungicides.

45. What is the concept of "refuge strategy" in fungicide resistance management?

The refuge strategy involves planting a portion of the crop with a non-resistant variety to maintain a susceptible pest population, which can mate with resistant individuals and dilute resistance genes.

46. How can education and awareness programs contribute to fungicide resistance management?

Education programs can inform farmers and agricultural professionals about the importance of responsible fungicide use and the potential consequences of resistance, encouraging more sustainable practices.

47. What are some examples of fungicides that have seen resistance develop in various regions?

Examples include resistance to triazoles in wheat pathogens, strobilurins in rice diseases, and QoI inhibitors in grapevine diseases.

48. What is the primary mission of FRAC (Fungicide Resistance Action Committee)?

The primary mission of FRAC is to prolong the effectiveness of fungicides by providing guidance and recommendations for the sustainable use of fungicides to prevent or delay the development of resistance in fungal pathogens.

49. How does FRAC categorize fungicides?

FRAC categorizes fungicides into different groups based on their mode of action, known as FRAC codes, to help users identify fungicides with similar mechanisms and avoid overuse of fungicides from the same group.

50 Why is it important for fungicides to be classified into FRAC groups?

Classifying fungicides into FRAC groups helps users diversify their fungicide applications, reducing the risk of resistance development, and making fungicide management more sustainable.

7

Disease Forecasting in Agriculture and Pathometry

A management tool for anticipating the onset of diseases in plants is called plant disease forecasting. This management method makes use of data on the past, present, and predicted weather in a particular area to forecast the onset and severity of disease in the future. In this sense, a plant disease forecasting system notifies farmers ahead of time whether or not to modify their crop protection strategies to ward against pests. The disease forecasting method lowers financial costs, farmer production loss, and negative environmental effects when it is timely and precisely forecasted. These systems ought to be prepared for diseases like these, which don't happen often but rather manifest destructively during opportune weather.

Requirements for Disease Forecasting

Systems for forecasting plant diseases are based on the "disease triangle," which describes how the environment, host, and pathogen interact. Therefore, information on these variables is necessary for the accurate forecasting of a disease outbreak in the future. Following questions and answers cover various aspects of disease forecasting and pathometry, including their impact on integrated disease management, environmental considerations, and specific challenges in different agricultural settings.

1. Host factors

A host can be susceptible or resistant. The stages of a host plant also determine the development of diseases. Some pathogen attack in seedling stages, while other infect when plant is grown-up. The outbreak of a disease also depends upon the population of plant. Densely populated plantation favors, while scattered plantation suppress the disease epidemics.

2. Pathogen factors

Based on previous history or survey data the presence of pathogen determine the disease epidemics. Data on amount of inoculum, their germination, dispersal, incubation period, sporulation, perenating stages are essential for disease forecasting system for accurate prediction of disease.

3. Environment factors

Environmental factors play a very critical role in interaction between a host and pathogen.Current data on temperature, relative humidity, direction and speed of wind are utilized as environmental factors in determining the outbreak of a plant disease.

Methods of Disease Forecasting

1. Forecasting based on primary inoculums

In this method, presence of primary inoculum, their density and viability is tested in the planting material, soil or in the air. Planting materials are randomly tested by different testing methods and recommendations are made for the chemical treatment of seed. Diseases, like Smut of wheat, ergot of pearl millet can be tested easily. In soil, presence and density of pathogens are tested by culturing them on specific culture medium. In air, spore of the pathogens are determined through the spore trap method.

2. Forecasting based on weather conditions

In this method, different parameters of weather conditions during and between the crop seasons are considered. These parameters include, temperature, relative humidity, rainfall, wind direction, light, etc. Weather conditions above the crops and of soil is also measured.

3. Forecasting based on correlative information

In this method, data of several years on weather is collected and correlated with the occurrence and intensity of the diseases. On basis of correlation disease forecasting is done. On basis of correlative information forecasting of barley powdery mildew and fire blight of apple have been made.

4. Computer-based disease forecasting models

These models works by processing the data on above mentioned factors and warn about the outbreak and severity of a diseases in near future. Among the computer-based models, EPIDEM was developed in 1969 for early bight of potato and tomato caused by *Alternaria solani*. Since then following models have been established to simulate the disease epidemics.

Forecast system	Diseases
EPIDEM	Early blight of potato and tomato caused by *Alternaria solani*
TOMCAST/FAST	Early blight of potato and tomato caused by *Alternaria solani*
MYCOS	Mycosphaerella blight of Chrysanthemum
EPIVEN	Apple scab caused by *Venturia inaequalis*
PLASMO	Downy mildew of grapevine caused by *Plasmopara viticola*
EPICORN	Southern corn leaf blight caused by *Helminthosporium maydis*
BLIGHTCAST	Late blight of potato caused by *Phytophthora infestans*
INDO-BLIGHT CAST	Late blight of potato caused by *Phytophthora infestans*
PHYTOPROG	Late blight of potato caused by *Phytophthora infestans*
CERCOS	Cercospora blight of celery
EPIDEMIC	Designed for stripe rust of wheat, but could be modified for other diseases
MARYBLIGHT	Fire blight on apple caused by *Erwinia amylovora*

Examples of disease forecasting

1. Late blight of potato

Late blight of potato is forecasted after occurrence of 7 to 14 days of blight favorable days. Blight favorable days are when, 5 day average temperature is 25.5°C and the total rainfall for the last 10 days is more than 3.0 cm. Nowadays, computerized models, such as, BLIGHTCAST, Indo-Blight Cast and Phytoprog are available in different parts of the world, which tells the farmers about the outbreak of late blight of potato in advance. Blightcast is operated by Syngenta, U.K. Indo-blightcast is operated by CPRI and AICRP, Shimla in collaboration with Agromet Division of Indian Meteorological Department, New Delhi. Phytoprog is the forecasting model used in West Germany.

2. Rice blast

Rice blast caused by *Pyricularia oryzae* is predicted on basis of correlative information method. The disease is predicted when, minimum night temperature range between 20 to 26 °C in association with 90 % or above relative humidity.

3. What is disease forecasting in agriculture, and why is it important?

Disease forecasting in agriculture is the process of predicting the occurrence, spread, and severity of crop diseases before they actually manifest in the field. It involves the use of scientific methods, data analysis, and modeling techniques to anticipate when and where diseases are likely to occur, as well as their potential impact on crops. Disease forecasting is essential for several reasons

1. Early Intervention Disease forecasting allows farmers to take proactive measures to manage and mitigate the impact of diseases on their crops. By identifying potential disease outbreaks in advance, farmers can apply preventive measures before the disease becomes widespread.
2. Minimized Crop Losses Timely disease forecasts help minimize crop losses by enabling farmers to implement control strategies, such as adjusting planting schedules, applying pesticides or fungicides, or using disease-resistant crop varieties.
3. Optimized Resource Use Disease forecasting helps farmers optimize the use of resources, such as pesticides and water. Instead of applying treatments blindly, they can target their interventions where and when they are needed most, reducing unnecessary chemical use.
4. Cost Reduction By preventing or reducing disease outbreaks, disease forecasting can lead to cost savings for farmers. This includes savings on the cost of chemicals, labor, and potential crop replanting.
5. Environmental Impact Disease forecasting supports environmentally sustainable farming practices by minimizing the use of agrochemicals and reducing the environmental impact of farming operations.
6. Food Security Effective disease management through forecasting contributes to food security by ensuring that crop yields are not compromised by disease-related losses. It helps ensure a stable food supply.
7. Research and Development- Disease forecasting drives research and development efforts to better understand crop diseases and develop improved disease-resistant crop varieties and management strategies.

8. Crop Quality Disease outbreaks can not only reduce crop yields but also affect the quality of harvested produce. Disease forecasting helps maintain higher crop quality, which is crucial for marketability.
9. Farmers' Income By protecting crops from disease, forecasting can help farmers maintain or increase their income. Healthy crops typically command better prices in the market.
10. Data-Driven Decision-Making Disease forecasting relies on data collection and analysis. It promotes a data-driven approach to agriculture, helping farmers make more informed decisions about disease management.

In summary, disease forecasting in agriculture is vital for preventing and managing crop diseases effectively. It empowers farmers to respond proactively to potential threats, reduces crop losses, conserves resources, and contributes to the overall sustainability and profitability of agricultural operations.

4. How does weather influence disease forecasting in agriculture?

Weather plays a significant role in disease forecasting in agriculture as it directly affects the development, spread, and severity of many crop diseases. Here's how weather influences disease forecasting

1. **Temperature:** Temperature is a critical factor in disease development. Many pathogens have temperature ranges in which they thrive. Disease forecasting models take into account temperature data to predict when conditions will be favorable for disease development. For example, certain fungal diseases are more likely to occur during warm, humid conditions, while others may be favored by cooler temperatures.
2. **Humidity:** Humidity levels can influence disease development, particularly for fungal diseases. High humidity provides the moisture necessary for fungal spore germination and growth. Disease forecasts often include humidity data to assess the risk of diseases like powdery mildew and downy mildew.
3. **Rainfall:** Rainfall can impact disease forecasting in several ways. Excessive rainfall can promote the spread of waterborne diseases, while prolonged wet conditions can create a favorable environment for fungal pathogens. Conversely, drought conditions can stress plants and make them more susceptible to diseases. Forecasting models consider both excessive rainfall and drought as risk factors.
4. **Wind:** Wind can facilitate the spread of pathogens by carrying spores or insects that transmit diseases from one plant to another. Disease forecasting models may incorporate wind patterns to predict disease spread, particularly for vector-borne diseases.
5. **Solar Radiation:** Solar radiation, including sunlight, can have both positive and negative effects on diseases. Some pathogens are sensitive to UV radiation, and sunlight can help reduce disease severity. Disease forecasting models may consider solar radiation levels to predict disease risk.
6. **Freezing Conditions:** Cold temperatures can affect disease forecasting, especially for diseases that are sensitive to freezing. Frost or freeze events can kill pathogens or slow their growth, reducing disease risk. Conversely, mild

winter conditions can allow pathogens to survive and increase disease pressure in the following growing season.

7. **Seasonal Changes:** Disease forecasts often consider seasonal changes, as different diseases are more prevalent during specific seasons. For example, late blight in tomatoes and potatoes is more common in late summer and early fall, so forecasts during these periods are crucial.
8. **Long-Term Climate Patterns:** Long-term climate patterns, such as *El Niño* and *La Niña* events, can influence weather patterns and, subsequently, disease risk. These climate phenomena can lead to unusual weather conditions that may affect disease forecasting.

Disease forecasting models combine historical weather data with current and forecasted weather conditions to predict disease risk. These models can provide farmers with valuable information about when and where to implement disease management strategies, such as applying fungicides or adjusting planting dates, to reduce the impact of diseases on their crops. Weather data is an essential component of disease forecasting, helping farmers make informed decisions to protect their agricultural investments.

5. Can you explain the concept of biological models in disease forecasting?

Biological models in disease forecasting are mathematical or computational models that incorporate biological factors and processes to predict the occurrence, spread, and severity of diseases in agriculture. These models are designed to provide a more holistic understanding of disease dynamics by taking into account the biological aspects of both the pathogens (disease-causing organisms) and the host plants.

Here are the key components and concepts associated with biological models in disease forecasting

1. **Pathogen Life Cycle:** Biological models often consider the life cycle of the pathogen responsible for the disease. This includes understanding when and where the pathogen reproduces, how it spreads, and how it interacts with the host plant. For example, models may incorporate information about pathogen spore production, dispersal mechanisms, and infection processes.
2. **Host Plant Susceptibility:** These models take into account the susceptibility of the host plant to the disease. Different plant varieties or species may have varying levels of resistance or susceptibility to specific pathogens. Biological models use this information to predict which plants are most at risk.
3. **Infection Periods:** Biological models identify critical periods during the growing season when environmental conditions are conducive to disease development. For example, models may pinpoint times when temperature, humidity, and rainfall favour pathogen growth and infection.
4. **Epidemiological Factors:** These models consider epidemiological factors, such as the rate of pathogen reproduction, the density of host plants, and the presence of vectors (organisms that transmit diseases). These factors influence disease spread and severity.

5. **Disease Progression:** Biological models simulate how the disease progresses over time and space. They can predict how quickly the disease will spread from an initial infection point and how severe the outbreak is likely to become.
6. **Control Strategies:** Some biological models may also incorporate information about disease control strategies, such as the application of fungicides, resistant crop varieties, or cultural practices like crop rotation. By considering these factors, the models can help optimize control measures.
7. **Validation:** Biological models are validated using field data to ensure their accuracy and reliability. Field observations and data on disease incidence and severity are used to assess the model's performance and make any necessary adjustments.
8. **Risk Assessment:** These models provide a quantitative assessment of disease risk, allowing farmers to make informed decisions about disease management strategies. For example, they may indicate when it is most critical to apply fungicides or implement other control measures.

Overall, biological models in disease forecasting offer a more comprehensive understanding of disease dynamics by integrating biological knowledge of pathogens, host plants, and the environment. They can be powerful tools for farmers, agricultural researchers, and extension services to make informed decisions to prevent and manage crop diseases effectively while minimizing the use of chemical inputs.

6. What role does remote sensing technology play in disease forecasting?

Remote sensing technology plays a crucial role in disease forecasting in agriculture by providing timely and detailed information about crop health, environmental conditions, and disease-related stress factors. Here's how remote sensing contributes to disease forecasting

1. **Early Detection** Remote sensing allows for the early detection of disease symptoms or stress factors in crops. Sensors on satellites, drones, or ground-based instruments can capture images and data that reveal subtle changes in crop color, texture, or temperature, which can be indicative of disease presence.
2. **Monitoring Crop Health** Remote sensing technology provides continuous and non-invasive monitoring of crop health over large areas. It helps identify areas of the field that are stressed or affected by diseases, even before visible symptoms appear.
3. **Spatial Mapping** Remote sensing enables the creation of detailed spatial maps showing disease distribution within a field or across an entire agricultural region. These maps can be used to target disease management efforts more precisely.
4. **Environmental Data** Remote sensing platforms can collect environmental data, such as temperature, humidity, and rainfall, which are essential factors in disease development. Integrating these environmental data into disease forecasting models improves their accuracy.
5. **Multi-Spectral and Hyperspectral Imaging** Advanced sensors can capture data in multiple spectral bands, allowing for the detection of subtle changes in plant physiology and stress responses. Hyperspectral imaging provides

even higher spectral resolution, enabling the differentiation of specific disease symptoms.

6. **Timely Updates** Remote sensing technology can provide frequent updates, sometimes on a daily or weekly basis, allowing for real-time or near-real-time disease monitoring. This timeliness is crucial for disease forecasting and rapid response.
7. **Large-Scale Coverage** Remote sensing can cover vast agricultural areas efficiently, providing a comprehensive view of disease distribution and severity across regions. This scalability is essential for monitoring large-scale crop production.
8. **Data Integration** Remote sensing data can be integrated into disease forecasting models, enhancing their predictive capabilities. Combining remote sensing with other data sources, such as weather data and historical disease records, improves the accuracy of forecasts.
9. **Resource Allocation** By pinpointing areas with higher disease risk, remote sensing helps farmers allocate resources more efficiently. They can target interventions, such as pesticide applications or irrigation, to the areas that need them most.
10. **Cost-Efficiency** Remote sensing technology can reduce the cost of disease monitoring compared to traditional field scouting, as it covers large areas quickly and can be automated.

In summary, remote sensing technology provides valuable data and imagery that enhance the accuracy and effectiveness of disease forecasting in agriculture. It enables early detection, precise monitoring, and informed decision-making, ultimately aiding farmers in managing crop diseases and reducing yield losses.

7. How can historical disease data improve disease forecasting accuracy?

Historical disease data can improve disease forecasting accuracy by providing valuable insights into past disease occurrences and patterns. By analyzing historical data, forecasters can

- **Identify Trends:** Recognize long-term trends in disease occurrence, such as seasonal patterns or cyclic outbreaks.
- **Understand Environmental Factors:** Determine how weather, climate, and other environmental variables correlate with disease outbreaks.

  Establish Baselines: Establish a baseline for normal disease levels, helping detect deviations and abnormal disease events.
- **Predict Disease Timing:** Use historical timing data to anticipate when diseases are likely to occur in future seasons.
- **Validate Models:** Test and validate disease forecasting models using historical data to assess their accuracy and reliability.

Incorporating historical disease data allows forecasters to make more informed and accurate predictions, aiding in effective disease management and crop protection.

8. What is the significance of early warning systems in disease forecasting?

Early warning systems notify farmers in advance of impending disease outbreaks, enabling timely interventions to mitigate crop losses.

Early warning systems in disease forecasting are of paramount significance in agriculture for several compelling reasons

1. **Timely Action:** Early warning systems provide farmers and agricultural stakeholders with advance notice of impending disease outbreaks. This timely information allows them to take proactive measures to protect crops before significant damage occurs.
2. **Reduced Crop Losses:** By enabling swift response to disease threats, early warning systems can substantially reduce crop losses. Farmers can implement control measures such as pesticide application, disease-resistant crop varieties, or cultural practices to mitigate the impact of diseases.
3. **Optimized Resource Use:** Early warnings allow for efficient resource allocation. Farmers can apply treatments or interventions precisely where needed, minimizing the unnecessary use of chemicals and resources.
4. **Economic Benefits:** Reduced crop losses and optimized resource use translate into financial savings for farmers. Early warning systems help safeguard farmers' income and profitability.
5. **Sustainability:** By preventing or minimizing disease outbreaks, early warning systems support sustainable agriculture. Reduced reliance on chemical pesticides and increased crop yield stability contribute to long-term agricultural sustainability.
6. **Food Security:** Early warnings play a vital role in ensuring food security. They help maintain stable crop production, reducing the risk of food shortages and price volatility in the market.
7. **Research and Development:** Early warning systems stimulate research and development efforts in disease-resistant crop varieties, improved disease management practices, and more effective control strategies.
8. **Capacity Building:** Implementing early warning systems encourages capacity building among farmers and agricultural extension services. It promotes education and awareness about disease prevention and management.
9. **Climate Adaptation:** As climate change affects disease dynamics, early warning systems aid in climate adaptation strategies. They provide data on changing disease patterns, allowing farmers to adjust their practices accordingly.

9. How do machine learning and artificial intelligence contribute to disease forecasting?

Machine learning (ML) and artificial intelligence (AI) offer substantial contributions to disease forecasting in agriculture through their data-driven and adaptive capabilities. Here are several key ways in which ML and AI enhance disease forecasting

1. **Data Integration:** ML and AI algorithms can integrate diverse data sources, including historical disease records, weather data, satellite imagery, and soil information. This comprehensive dataset enables a more holistic understanding of disease dynamics.

2. **Pattern Recognition:** ML and AI excel at recognizing complex patterns and correlations within vast datasets. They can identify subtle relationships between environmental factors and disease outbreaks that may elude traditional statistical models.
3. **Predictive Models:** These technologies build predictive models that continuously learn from new data, allowing for real-time disease forecasts. As more data becomes available, these models become more accurate and responsive.
4. **Early Detection:** ML and AI can identify early indicators of disease outbreaks, sometimes before visible symptoms manifest in crops. This early detection is crucial for timely intervention.
5. **Precision Agriculture:** ML and AI enable precise and targeted disease forecasts at the field level. Farmers can receive recommendations specific to their individual fields, optimizing resource allocation and treatment strategies.
6. **Adaptive Learning:** These technologies can adapt to changing conditions, such as evolving disease strains or climate shifts, ensuring that disease forecasts remain relevant and effective over time.
7. **Efficiency:** ML and AI automate data analysis and model training, significantly reducing the time and effort required for disease forecasting compared to manual methods.
8. **Scalability:** ML and AI can scale to monitor large agricultural regions, providing valuable insights for regional and national disease management strategies.
9. **Enhanced Decision Support:** By providing actionable insights and risk assessments, ML and AI empower farmers and agricultural experts to make informed decisions regarding disease management, resource allocation, and crop protection measures.

In summary, machine learning and artificial intelligence revolutionize disease forecasting by leveraging advanced algorithms to process vast and diverse datasets, recognize intricate patterns, and deliver timely, precise, and adaptive forecasts. Their contributions empower agriculture stakeholders to proactively protect crops, reduce yield losses, and promote sustainable and efficient farming practices.

10. What challenges are associated with predicting the spread of vector-borne diseases in agriculture?

Predicting the spread of vector-borne diseases in agriculture presents several challenges due to the complexity of the interactions involved. Some of the key challenges include

1. **Vector Behavior and Biology:** The behavior and biology of disease vectors, such as insects or nematodes, can be intricate and difficult to predict. Factors like feeding habits, reproduction rates, and host preferences influence the vector's ability to transmit diseases.
2. **Environmental Variability:** Environmental conditions, including temperature, humidity, and rainfall, impact vector populations and disease transmission. Climate variability and changing weather patterns make it challenging to anticipate disease dynamics accurately.

3. **Vector Mobility:** Vectors can move across large distances, sometimes faster than diseases spread. Predicting their movement patterns and the potential introduction of diseases to new areas is complex.
4. **Host-Vector Interactions:** Understanding the interactions between vectors and host plants is crucial. Different crop varieties may have varying levels of susceptibility, and vector preferences can influence disease transmission.
5. **Emerging Diseases:** The emergence of new diseases or the introduction of exotic vectors can be unpredictable. Predictive models often rely on historical data, which may not account for novel pathogens or vectors.
6. **Resistance Development:** The development of resistance in disease vectors to control measures, such as pesticides, can impact disease spread. Predicting the emergence of resistant strains is challenging.
7. **Data Availability:** Accurate data on vector populations, disease prevalence, and environmental conditions are essential for modeling disease spread. Data gaps or inconsistencies can hinder forecasting efforts.
8. **Complexity of Models:** Modeling vector-borne diseases involves complex mathematical and computational models that require extensive data and expertise. Model validation and calibration can be challenging.
9. **Human and Cultural Factors:** Human behavior, such as agricultural practices and land use changes, can influence vector populations and disease spread. Cultural practices may also affect the likelihood of disease outbreaks.
10. **Interdisciplinary Collaboration:** Effective disease prediction often requires collaboration between entomologists, plant pathologists, climatologists, and data scientists. Bridging gaps between disciplines can be a logistical challenge.
11. **Resource Constraints:** Developing and maintaining comprehensive disease forecasting systems can be resource-intensive. Many regions, especially in developing countries, may lack the necessary infrastructure and funding for robust disease prediction.

Despite these challenges, ongoing research and advancements in modeling techniques, data collection methods, and interdisciplinary collaboration continue to improve our ability to predict and manage vector-borne diseases in agriculture.

11. How can disease forecasting be tailored to different crops and regions?

Disease forecasting can be customized for diverse crops and regions by adapting models, data sources, and strategies to specific agricultural contexts. This involves considering factors such as crop types, local climate, soil conditions, and historical disease patterns. Customization also entails involving local experts, utilizing precision agriculture technologies, and providing alerts and recommendations in local languages through accessible channels. By tailoring disease forecasting, farmers can receive accurate, relevant information to make informed decisions and implement effective disease management practices that align with the unique challenges and conditions of their crops and regions.

12. What are the potential benefits of disease forecasting for sustainable agriculture?

Disease forecasting offers several potential benefits for sustainable agriculture

- **Reduced Crop Losses:** Early disease detection and intervention enable farmers to minimize crop losses, ensuring more reliable harvests and food security.
- **Optimized Resource Use:** Disease forecasts help farmers apply pesticides and other inputs only when necessary, reducing chemical use, resource waste, and environmental impact.
- **Sustainable Practices:** Disease management guided by forecasting promotes eco-friendly practices, such as organic farming, integrated pest management, and the use of disease-resistant crop varieties.
- **Financial Savings:** Reduced crop losses and optimized resource use translate into cost savings for farmers, improving economic sustainability.
- **Preservation of Biodiversity:** Sustainable disease management practices protect beneficial organisms and contribute to biodiversity conservation.
- **Research and Innovation:** Disease forecasting encourages research into disease-resistant crops and innovative management strategies, fostering long-term agricultural sustainability.
- **Stable Food Supply:** Disease forecasting contributes to a stable food supply by minimizing yield fluctuations due to diseases, which is vital for food security.
- **Environmental Protection:** Reduced chemical use and environmentally friendly practices contribute to the protection of soil, water, and ecosystems.
- **Climate Resilience:** Disease forecasts can help farmers adapt to changing climate conditions and anticipate disease shifts in response to climate change.
- **Data-Driven Decision-Making:** Disease forecasting promotes data-driven decision-making in agriculture, leading to more informed and sustainable practices.

In summary, disease forecasting is a powerful tool that supports sustainable agriculture by reducing crop losses, optimizing resource use, fostering eco-friendly practices, and contributing to food security and environmental protection.

Pathometry in Agriculture

13. What is pathometry, and how does it contribute to disease management in agriculture?

Pathometry, also known as plant pathology, is the scientific discipline focused on the study of plant diseases. It encompasses the identification, understanding, and management of diseases that affect crops, trees, and other plant species in agriculture. Pathometry contributes significantly to disease management in agriculture in several ways

1. **Disease Identification:** Plant pathologists identify and characterize diseases, pinpointing the pathogens responsible and their modes of infection. This knowledge is fundamental for recognizing and addressing specific diseases.

2. **Disease Diagnosis:** Pathologists diagnose diseases through laboratory tests, microscopic examinations, and field observations. Accurate diagnosis is crucial for implementing appropriate control measures.
3. **Resistance Breeding:** Pathometry aids in breeding crop varieties with inherent resistance to diseases, reducing the reliance on chemical treatments and promoting sustainable agriculture.
4. **Disease Monitoring:** Plant pathologists monitor disease outbreaks, providing real-time data on disease prevalence, severity, and distribution, which guides management decisions.
5. **Integrated Pest Management (IPM):** Pathometry contributes to IPM strategies by offering insights into disease dynamics and integrating multiple approaches to disease control, including biological control and cultural practices.
6. **Research and Innovation:** Pathological research informs the development of new disease management techniques, such as Bio control agents, disease-resistant crop varieties, and innovative cultural practices.
7. **Educational Outreach:** Pathologists collaborate with extension services to educate farmers and agricultural stakeholders about disease prevention, management, and sustainable practices.

14. How are pathogens responsible for crop diseases identified and classified?

Pathogens causing crop diseases are identified and classified through a systematic process that combines various techniques

1. **Symptomatology:** Initial identification involves observing and analyzing the symptoms exhibited by infected plants. These symptoms, such as leaf discoloration, lesions, or wilting, offer clues about the nature of the pathogen.
2. **Morphological Characteristics:** Pathogens can be visually characterized based on their size, shape, color, and the presence of unique structures like spores, hyphae, or bacterial colonies.
3. **Cultural Characteristics:** Pathogens can be grown in the laboratory on specific media. Their growth characteristics, such as colony appearance and texture, are examined and compared to reference materials.
4. **Microscopy:** Microscopic examination of pathogen structures is crucial. Fungi and bacteria, for example, have distinct microscopic features that aid in differentiation.
5. **Molecular Techniques:** DNA-based methods like PCR and DNA sequencing target specific genetic markers unique to the pathogen, providing precise identification.
6. **Host Range:** Understanding the range of plants a pathogen can infect helps narrow down its identity.
7. **Serological Tests:** These tests detect specific antibodies or antigens associated with pathogens, aiding in identification.
8. **Pathogenicity Tests:** Inoculating healthy plants with the suspected pathogen confirms its ability to cause disease.

9. **Biochemical Tests:** Some pathogens are identified through tests that assess their metabolic processes or enzyme activities.
10. **Genomic Analysis** Advanced techniques such as whole-genome sequencing provide comprehensive genetic information for precise identification.

These methods collectively enable scientists and plant pathologists to accurately identify and classify pathogens responsible for crop diseases, guiding effective disease management strategies and crop protection measures.

15. What are the key diagnostic tools used in pathometry?

Key diagnostic tools in pathometry, the field of plant pathology, include

1. **Microscopy:** Microscopes are used to examine pathogen structures like spores, hyphae, or bacterial cells.
2. **Cultural Characteristics:** Pathogens are grown in lab cultures, and their growth patterns are observed.
3. **Serological Tests:** These detect specific antibodies or antigens associated with pathogens.
4. **Molecular Techniques:** PCR and DNA sequencing identify pathogens based on genetic markers.
5. **Host Range Studies:** Understanding which plants a pathogen can infect is informative.
6. **Biochemical Tests:** Some pathogens are identified based on metabolic processes or enzyme activities.
7. **Pathogenicity Tests** Inoculating healthy plants with a suspected pathogen confirms its disease-causing ability.
8. **Genomic Analysis:** Whole-genome sequencing and comparative genomics provide detailed genetic information for identification.
9. **Symptomatology:** Analyzing disease symptoms on plants provides initial clues about the pathogen.

These tools aid in accurate identification, classification, and diagnosis of plant diseases, supporting effective disease management in agriculture.

16. How do agricultural experts diagnose diseases based on plant symptoms?

Agricultural experts diagnose diseases based on plant symptoms through a systematic process. They

1. **Observe Symptoms** Carefully examine the plant for visible signs like leaf discoloration, lesions, or wilting.
2. **Symptom Patterns** Analyze the pattern and distribution of symptoms on the plant, noting if they are localized or systemic.
3. **Compare to Reference** Compare observed symptoms to reference materials, such as disease guides or databases, to identify potential pathogens.
4. **Environmental Factors** Consider environmental conditions like weather, soil, and moisture levels, as these can influence symptom expression.

5. **Laboratory Confirmation** In cases of uncertainty, experts may conduct laboratory tests, including microscopy, serological tests, or molecular techniques, to confirm the presence of specific pathogens.
6. **Experience** Rely on their experience and expertise to make informed diagnoses, considering factors like the plant's history and region.

By combining these methods, agricultural experts accurately diagnose plant diseases, enabling targeted and effective disease management strategies.

17. Can you explain the role of molecular diagnostic methods in pathometry?

Molecular diagnostic methods are pivotal in pathometry, as they offer precise identification of pathogens responsible for plant diseases. These techniques involve the analysis of the pathogen's genetic material, enabling accurate and rapid diagnoses. Key roles include

1. **Specific Identification** Molecular methods target unique genetic markers, allowing for precise identification of pathogens at the species or strain level.
2. **Early Detection** These methods can detect pathogens even before visible symptoms appear in plants, facilitating timely intervention.
3. **Disease Monitoring** Molecular tools aid in monitoring the prevalence and distribution of pathogens in agricultural settings.
4. **Pathogen Characterization** Detailed genetic information helps researchers understand the biology, behavior, and potential for virulence or resistance of pathogens.
5. **Phylogenetic Studies** Molecular data contribute to phylogenetic studies, shedding light on the evolutionary relationships among pathogens.
6. **Customized Solutions** Molecular diagnostics inform the development of tailored disease management strategies, such as resistant crop varieties or targeted treatments, promoting sustainable agriculture.

18. How do cultural practices impact disease management in agriculture?

Cultural practices significantly influence disease management in agriculture by altering the crop environment and reducing disease pressure. Key impacts include

1. **Crop Rotation** Rotating crops disrupts disease life cycles, reducing pathogen buildup in the soil.
2. **Plant Spacing** Proper plant spacing enhances air circulation, reducing humidity and the risk of foliar diseases.
3. **Sanitation** Removing and destroying infected plant material prevents pathogen overwintering.
4. **Resistant Varieties** Planting disease-resistant crop varieties reduces susceptibility.
5. **Seed Selection** Using disease-free seeds and propagative materials minimizes initial pathogen introduction.
6. **Timing** Timely planting and harvesting can avoid disease-conducive conditions.
7. **Fertilization and Irrigation** Balanced nutrient management and appropriate irrigation prevent stress that makes plants more susceptible to diseases.

These practices, when integrated into a holistic disease management strategy, enhance crop health and reduce the need for chemical interventions.

19. What is the significance of genetic resistance in disease management?

Genetic resistance is of paramount significance in disease management for several reasons

1. **Reduced Chemical Use** Resistant crop varieties require fewer chemical treatments, minimizing environmental impact and lowering production costs.
2. **Sustainability** Genetic resistance promotes sustainable agriculture by reducing reliance on synthetic pesticides, fostering eco-friendly farming practices.
3. **Stability** Resistant crops provide stable yields by minimizing losses to diseases, ensuring food security and economic stability.
4. **Long-term Solution** Genetic resistance offers a long-lasting solution, as it is not compromised by pathogen evolution or resistance development.
5. **Efficiency** It streamlines disease management efforts, making them more efficient and cost-effective.
6. **Biodiversity** Resistant crop varieties protect biodiversity by reducing the need for broad-spectrum pesticides that can harm non-target organisms.

Overall, genetic resistance plays a pivotal role in sustainable, environmentally friendly, and economically viable disease management in agriculture.

20. How do Bio control agents contribute to disease management?

Bio control agents contribute to disease management in agriculture by harnessing natural predators or antagonists to control plant pathogens. Key roles include

1. **Biological Suppression** Bio control agents, such as beneficial insects or microorganisms, prey on or compete with pathogens, reducing their populations.
2. **Disease Prevention** Some Bio control agents stimulate the plant's natural defenses, making them more resistant to diseases.
3. **Selective Targeting** Bio control is often specific to the pathogen, minimizing harm to non-target organisms and reducing environmental impacts.
4. **Sustainable Approach** It promotes eco-friendly practices, reducing reliance on chemical pesticides and preserving beneficial organisms.
5. **Long-term Efficacy** Bio control agents can provide long-lasting disease protection, especially when integrated into holistic pest management programs.

Overall, Bio control agents offer an environmentally sustainable and effective alternative to chemical pesticides for disease management in agriculture.

21. What are some challenges in chemical pesticide use for disease management?

Chemical pesticide use in disease management faces several challenges

1. **Environmental Impact** Pesticides can harm non-target species, pollinators, and aquatic ecosystems, leading to ecological imbalances.
2. **Pesticide Resistance** Pathogens can develop resistance to pesticides over time, rendering them less effective.
3. **Health Concerns** Pesticide exposure poses health risks to farmworkers, consumers, and nearby communities.

4. **Residue Buildup** Pesticide residues may accumulate in soil, water, and food, affecting human health and the environment.
5. **Broad-Spectrum Effects** Chemical pesticides often kill beneficial organisms, disrupting natural pest control.
6. **Regulatory Challenges** Stringent regulations govern pesticide use, increasing compliance and cost burdens.
7. **Environmental Persistence** Some pesticides persist in the environment for extended periods, posing long-term risks.

Addressing these challenges requires a shift toward sustainable, integrated pest management practices that reduce pesticide reliance while effectively managing crop diseases.

22. How does pathometry research contribute to the development of sustainable disease management practices?

Pathometry research contributes to sustainable disease management practices in agriculture by providing valuable insights into disease dynamics, pathogen behavior, and control strategies. This research

1. **Identifies Resistant Varieties** Pathometry helps breeders identify and develop crop varieties resistant to specific diseases, reducing the need for chemical treatments.
2. **Optimizes Timing** It informs farmers about the optimal timing for planting, harvesting, and disease control measures.
3. **Enhances Bio control** Pathometry studies identify potential bio control agents and strategies for biological disease management.
4. **Guides IPM** Pathometry data supports integrated pest management (IPM) programs, which combine cultural, biological, and chemical methods for sustainable disease control.
5. **Improves Diagnostic Tools** Research enhances disease diagnostics, facilitating early detection and targeted interventions.

Overall, pathometry research enhances disease management strategies, promoting sustainability, reducing environmental impact, and ensuring long-term crop health.

Integrated Disease Management

23. How does disease forecasting integrate with the concept of integrated pest management (IPM)?

Disease forecasting integrates seamlessly with the concept of Integrated Pest Management (IPM) by providing timely and data-driven insights into disease dynamics. Here's how

1. **Early Warning** Disease forecasts offer early warnings about potential outbreaks, allowing IPM practitioners to take proactive measures.
2. **Targeted Interventions** IPM strategies can be customized based on disease forecasts, optimizing the timing and choice of control measures.
3. **Reduced Chemical Use** Disease forecasts help reduce reliance on chemical pesticides by promoting precision in their application.

4. **Resource Efficiency** IPM, guided by disease forecasts, optimizes resource use, minimizing environmental impact.
5. **Sustainable Practices** The integration of disease forecasting aligns with IPM's focus on sustainable and environmentally friendly pest and disease management.

24. What are the ecological implications of disease management practices in agriculture?

Disease management practices in agriculture have several ecological implications

1. **Biodiversity Impact** Chemical pesticides can harm non-target organisms, including beneficial insects and microbes, impacting ecosystem diversity.
2. **Pollution** Pesticide runoff can contaminate water bodies, harming aquatic ecosystems and affecting water quality.
3. **Resistance Development** Frequent pesticide use can lead to the development of resistant pathogens, requiring more potent chemicals.
4. **Disruption of Natural Controls** Pesticides can disrupt natural predation and parasitism, upsetting the balance between pests and their natural enemies.
5. **Soil Health** Some pesticides can harm beneficial soil organisms, affecting soil health and nutrient cycling.
6. **Food Chain Effects** Pesticides can accumulate through the food chain, affecting higher trophic levels.

Balancing effective disease management with ecological sustainability requires adopting practices such as integrated pest management (IPM), Bio control, and organic farming to reduce the environmental impact of disease control measures.

25. How does sustainable disease management differ from conventional disease management practices?

Sustainable disease management differs from conventional practices by prioritizing environmentally friendly, long-term solutions

1. **Chemical vs. Non-Chemical** Sustainable management minimizes chemical pesticide use, favoring biological controls, resistant crop varieties, and cultural practices.
2. **Ecosystem Health** It emphasizes protecting ecosystem health, preserving biodiversity, and minimizing harm to non-target organisms.
3. **Resilience** Sustainable approaches promote disease-resistant crop varieties, reducing the reliance on chemicals and the risk of resistance development.
4. **Long-Term Focus** Conventional methods often prioritize short-term disease suppression, while sustainable practices aim for enduring solutions.
5. **Holistic Approach** Sustainable management integrates various strategies, like integrated pest management (IPM), to reduce environmental impact and promote long-term agricultural viability.

26. Can you explain the role of Bio control agents in disease management, and what are their advantages?

Bio control agents play a crucial role in disease management by harnessing natural predators or antagonists to control plant pathogens. Their advantages include

1. **Environmentally Friendly** Bio controls are eco-friendly and reduce the environmental impact of chemical pesticides.
2. **Specific Targeting** They are often species-specific, minimizing harm to non-target organisms.
3. **Reduced Resistance** Bio control agents pose lower risks of pathogen resistance development.
4. **Long-Lasting** Some Bio controls establish resident populations, offering enduring disease protection.
5. **Sustainable** Bio controls align with sustainable agriculture practices, reducing the reliance on synthetic pesticides.
6. **Safe** They are generally safe for humans, animals, and beneficial insects.
7. **Integrated Pest Management (IPM)** Bio controls are a key component of IPM strategies that promote holistic, balanced, and effective disease management.

27. How can crop rotation effectively be used as a disease management strategy?

Crop rotation is an effective disease management strategy that relies on alternating different plant species in specific sequences. Its benefits include

1. **Disease Break:** Crop rotation disrupts the life cycles of pathogens, reducing their buildup in the soil.
2. **Host Specificity:** Changing crops minimizes the host plants available to pathogens, preventing disease carryover.
3. **Nutrient Cycling:** Different crops have varying nutrient requirements, allowing for more balanced soil nutrient levels.
4. **Weed and Pest Suppression:** Rotation can deter specific weeds and pests associated with particular crops.
5. **Improved Soil Health:** Crop diversity promotes beneficial soil microorganisms and reduces soil-borne diseases.
6. **Reduced Chemical Use:** Fewer chemicals are needed for disease control in a well-managed rotation system, benefiting both the environment and farm economics.

28. What ethical considerations should be taken into account when implementing disease management strategies?

Ethical considerations in disease management strategies should encompass

1. **Environmental Stewardship** Prioritize practices that minimize harm to ecosystems, biodiversity, and natural resources.
2. **Human Health and Safety** Ensure the well-being of farmworkers, nearby communities, and consumers by using safe disease management methods.
3. **Sustainability** Promote practices that maintain agricultural viability over the long term while mitigating environmental impacts.
4. **Equity** Ensure that disease management strategies do not disproportionately affect disadvantaged or marginalized communities.
5. **Transparency** Communicate openly with stakeholders about chosen strategies, their potential risks, and benefits.

6. **Research and Monitoring** Continuously assess and adapt strategies based on their ecological and social impacts.

Ethical disease management balances short-term disease control with long-term sustainability and social responsibility.

29. Explain the importance of farmer education and extension services in disease management.

Farmer education and extension services are crucial in disease management

1. **Knowledge Transfer** They provide farmers with essential information on disease recognition, prevention, and control measures.
2. **Early Detection** Educated farmers can identify disease symptoms promptly, enabling timely intervention.
3. **Sustainable Practices** Extension services promote eco-friendly and sustainable disease management practices, reducing chemical use.
4. **Adaptation** Farmers are informed about emerging diseases and changing disease patterns, allowing them to adapt their strategies.
5. **Efficiency** Well-informed farmers use resources more efficiently, optimizing disease management efforts.
6. **Community Resilience** Educated farmers collectively contribute to stronger, disease-resistant farming communities.

By empowering farmers with knowledge and resources, education and extension services enhance disease management, benefiting both agricultural productivity and sustainability.

30. How do global trade and transportation networks influence the spread of plant diseases?

Global trade and transportation networks play a significant role in the spread of plant diseases

1. **Pathogen Movement** Infected plant materials, including seeds and produce, can carry pathogens across borders, introducing diseases to new regions.
2. **Vector Transport** Insects and pests that transmit diseases can hitch rides on cargo, facilitating the spread of pathogens.
3. **Climate Change** Trade networks can inadvertently introduce pathogens to areas with changing climates that are now conducive to their survival.
4. **Trade Volume** The high volume of global trade increases the likelihood of introducing and spreading diseases.
5. **Interconnections** As global networks expand, the interconnections of ecosystems makes it easier for pathogens to find suitable hosts and vectors, fueling disease spread.

Effective plant disease management often involves monitoring and regulating trade and transportation to mitigate these risks.

31. What are the economic implications of effective disease management for farmers and agriculture as a whole?

Effective disease management has several economic implications for farmers and agriculture

1. **Increased Yield and Profit** By reducing crop losses, farmers can achieve higher yields and increased profits.
2. **Cost Savings** Fewer resources are spent on pesticides, treatments, and replanting, reducing production costs.
3. **Market Access** Disease-free produce meets market standards, ensuring access to premium markets and better prices.
4. **Sustainability** Sustainable disease management practices promote long-term agricultural viability, safeguarding livelihoods and economic stability.
5. **Investment Attraction** Effective disease management makes agriculture an attractive investment, fostering growth and innovation.
6. **Food Security** Stable crop production ensures food security, reducing the need for costly food imports.

In summary, effective disease management contributes to economic prosperity, food security, and sustainability in agriculture.

32. How can stakeholders in agriculture collaborate to improve disease forecasting and management practices on a global scale?

Stakeholders in agriculture can collaborate globally to enhance disease forecasting and management

1. **Data Sharing** Share disease data, research, and surveillance information across borders to improve early detection.
2. **Research Partnerships** Foster international collaborations among research institutions to develop disease-resistant crop varieties and innovative management strategies.
3. **Capacity Building** Invest in training programs for farmers, extension workers, and scientists to build expertise in disease management.
4. **Technology Adoption** Promote the use of advanced technologies like remote sensing and artificial intelligence for accurate disease forecasting.
5. **Standardization** Develop global disease monitoring and reporting standards to ensure consistency and transparency.
6. **Policy Alignment** Harmonize international policies and regulations to facilitate the movement of disease-resistant crops and technologies.
7. **Early Warning Systems** Establish a global network of early warning systems to share alerts about emerging diseases.

Collaboration at this scale enhances the resilience of agriculture against disease threats, fosters sustainability, and ensures global food security.

Crop-Specific Considerations

33. What are the disease challenges specific to perennial crops, and how are they managed differently from annual crops?

Disease challenges in perennial crops, such as fruit trees and vines, differ from annual crops due to their longer lifespan. Perennial crop disease management involves

1. **Persistent Infections** Diseases like citrus greening can persist for years. Management requires continuous monitoring and control.
2. **Slow Crop Development** Perennial crops take years to mature, making early disease management crucial to avoid long-term losses.
3. **Pruning and Sanitation** Practices like pruning and proper sanitation are essential to remove infected plant parts and reduce pathogen reservoirs.
4. **Resistant Varieties** Breeding for disease-resistant perennial crop varieties is a long-term strategy.
5. **Integrated Approaches** Disease management integrates cultural practices, Bio controls, and reduced pesticide use to minimize environmental impact over the tree's lifespan.

Managing perennial crop diseases requires patience, ongoing vigilance, and tailored, sustainable approaches.

34. Can you explain the unique disease challenges in horticultural crops and the strategies used to address them?

Horticultural crops, such as fruits and vegetables, face distinct disease challenges

1. **Diversity** Horticultural crops encompass a wide range of species, each with its unique disease susceptibility.
2. **High Value** These crops often have high market value, making disease prevention crucial for economic reasons.
3. **Fungal Pathogens** Many horticultural crops are vulnerable to fungal diseases, necessitating vigilant management.
4. **Harvest Timing** Timing is critical, as diseases can affect the quality and shelf life of the harvested produce.

Strategies to address these challenges include using disease-resistant crop varieties, implementing integrated pest management (IPM) practices, practicing proper sanitation, and employing precision agriculture technologies for early detection and targeted interventions, all while minimizing environmental impacts.

35. How does disease forecasting contribute to the management of staple food crops that are critical for global food security?

Disease forecasting plays a vital role in managing staple food crops, which are essential for global food security

1. **Early Warning** Forecasting allows early detection of disease outbreaks, enabling timely intervention to prevent crop losses.
2. **Resource Allocation** Farmers can optimize resource use, such as water, fertilizers, and pesticides, based on forecasted disease risk.

3. **Resilience** Accurate forecasting helps breeders develop disease-resistant crop varieties, enhancing long-term food security.
4. **Market Stability** Stable production of staple crops ensures price stability and access to affordable food for consumers.
5. **Sustainability** Disease forecasting promotes sustainable farming practices by reducing the need for chemical inputs and preserving soil health.

Overall, disease forecasting is instrumental in safeguarding staple food crops, ensuring global food security, and mitigating hunger risks.

36. How do disease management strategies differ between monoculture and polyculture farming systems?

Disease management strategies differ between monoculture and polyculture farming systems

1. **Monoculture** In monoculture, a single crop is grown extensively. Disease management often relies heavily on chemical pesticides due to the higher disease risk in uniform plantings.
2. **Polyculture** Polyculture involves growing multiple crop species in proximity. Disease management is more diversified, as different crops can act as natural barriers, reducing disease spread. Crop rotation, companion planting, and biological control methods are more commonly employed.
3. **Crop Diversity** Polyculture naturally reduces disease pressure by disrupting pathogen life cycles and reducing host concentration.
4. **Sustainability** Polyculture promotes sustainable, eco-friendly farming by reducing pesticide use and enhancing soil health.

Overall, polyculture systems offer more diverse and holistic disease management strategies compared to monoculture systems.

37. Can you provide examples of region-specific disease forecasting and pathometry approaches?

Certainly, here are examples of region-specific disease forecasting and pathometry approaches

1. **Rice Blast in Southeast Asia** Southeast Asian countries employ disease forecasting models based on weather data to predict outbreaks of rice blast, a common fungal disease. This enables timely fungicide application.
2. **Citrus Canker in Florida, USA** Florida uses remote sensing technology to monitor citrus canker outbreaks. Pathometry studies help identify infected trees for targeted removal to prevent disease spread.
3. **Late Blight in Potato Farming in Europe** European nations utilize predictive models based on weather conditions to forecast late blight outbreaks in potato crops, guiding fungicide application and reducing environmental impact.

These region-specific approaches demonstrate the importance of tailoring disease management strategies to local conditions and crops.

38. How do temperature and humidity variations affect disease dynamics in different crops?

Temperature and humidity variations significantly impact disease dynamics in various crops

1. **Fungal Diseases** High humidity coupled with warm temperatures often promote the development of fungal diseases like powdery mildew and rust in many plants.
2. **Bacterial Diseases** Moist and warm conditions can lead to the proliferation of bacterial diseases, such as fire blight in apples.
3. **Viral Diseases** Temperature extremes can affect vector populations, influencing the transmission of viral diseases like leaf curl in tomatoes.
4. **Crop Specificity** Different crops have varying temperature and humidity preferences, affecting disease susceptibility.
5. **Geographical Variations** Regional climate differences can lead to varying disease patterns, necessitating tailored management strategies.

Understanding these dynamics is crucial for predicting disease outbreaks and implementing effective management practices.

39. What are the disease considerations specific to organic farming practices?

Disease management in organic farming presents unique challenges

1. **Limited Chemical Use** Organic farmers have restricted access to synthetic pesticides, necessitating alternative disease control methods.
2. **Crop Rotation** Organic practices often rely on crop rotation to disrupt disease cycles and prevent soil-borne diseases.
3. **Resistant Varieties** Disease-resistant crop varieties are preferred to reduce the reliance on chemical treatments.
4. **Bio control** Beneficial insects and microorganisms are used to control pests and pathogens.
5. **Sanitation** Proper sanitation practices help minimize disease spread.
6. **Compost** The use of disease-free compost is critical to avoid introducing pathogens.

Organic farming's emphasis on sustainability and natural practices requires innovative, integrated approaches to manage diseases effectively while complying with organic standards.

40. Explain how hydroponic and vertical farming systems impact disease management.

Hydroponic and vertical farming systems impact disease management in several ways

1. **Reduced Soil-Borne Diseases** Hydroponics eliminates soil, reducing the risk of soil-borne diseases that affect traditional farming.
2. **Controlled Environment** Vertical and hydroponic systems allow precise control over environmental factors like temperature and humidity, minimizing disease-friendly conditions.
3. **Isolation** These systems often involve isolated plant growth, limiting disease spread between plants.

4. **Monitoring** Advanced technology in these systems enables real-time disease monitoring, facilitating early detection and targeted interventions.
5. **Hygiene** Closed and controlled environments promote cleaner conditions, reducing disease pressure.

Overall, hydroponic and vertical farming systems offer more controlled and disease-resistant environments, improving disease management in intensive crop production.

41. How does precision agriculture technology contribute to disease management in modern farming?

Precision agriculture technology significantly enhances disease management in modern farming

1. **Remote Sensing** Drones and satellite imagery monitor crops for early disease symptoms, enabling quick responses.
2. **Data Analytics** Advanced algorithms process data from various sensors to predict disease outbreaks based on environmental conditions.
3. **Variable Rate Application** Precision equipment allows targeted application of treatments, reducing chemical use and minimizing environmental impact.
4. **Automated Monitoring** Sensor networks continuously monitor soil and plant health, alerting farmers to disease risks.
5. **GPS Guidance** GPS technology ensures precise application of disease management strategies, optimizing resource use.

By providing real-time data and enabling precise interventions, precision agriculture empowers farmers to proactively manage diseases, increasing crop yield and sustainability.

Integrated Disease Management

42. How does climate change adaptation influence disease management strategies in agriculture?

Climate change adaptation significantly influences disease management in agriculture

1. **Altered Disease Patterns** Changing climate conditions can shift the geographical distribution of diseases and the emergence of new pathogens.
2. **Timing and Intensity** Climate-related factors like temperature and humidity affect disease dynamics, requiring adjustments in disease control strategies.
3. **Resistant Varieties** Breeding for climate-resilient, disease-resistant crop varieties becomes crucial to adapt to changing disease pressures.
4. **Precision Farming** Technologies like remote sensing and data analytics help monitor climate-driven disease risks and enable precise, timely interventions.
5. **Integrated Approaches** Climate-adaptive disease management integrates various strategies to mitigate disease impacts, ensuring agricultural sustainability in a changing world.

43. What are the key biosecurity measures in place to prevent the introduction of new plant diseases to a region?

Key biosecurity measures to prevent the introduction of new plant diseases to a region include

1. **Quarantine** Inspecting and quarantining imported plant materials to detect and isolate potential pathogens.
2. **Regulations** Enforcing strict regulations on the movement of plants, seeds, and other agricultural materials.
3. **Surveillance** Regular monitoring and surveillance for early detection of new diseases or pathogens.
4. **Certification** Certification programs for disease-free plant materials and strict testing protocols.
5. **Border Control** Implementing rigorous border control measures, including inspections and disinfection.
6. **Public Awareness** Educating farmers, stakeholders, and the public on biosecurity measures to prevent accidental disease introduction.
7. **Research** Ongoing research to identify and understand emerging diseases and their vectors.

These measures collectively safeguard regions from the introduction and spread of new plant diseases.

Environmental and Regulatory Aspects

44. What role do government agencies play in disease monitoring and reporting in agriculture?

Government agencies play a pivotal role in disease monitoring and reporting in agriculture

1. **Surveillance** They establish and operate disease surveillance systems to detect outbreaks and monitor disease trends.
2. **Data Collection** Government agencies collect and compile data on disease occurrence, which aids in early detection.
3. **Reporting** These agencies disseminate timely and accurate information about disease outbreaks to farmers, researchers, and the public.
4. **Regulation** Government bodies enforce regulations on disease management, including quarantine measures and movement restrictions.
5. **Research** They often fund and conduct research to better understand diseases and develop effective control strategies.

Government agencies' involvement ensures a coordinated, proactive, and well-informed approach to disease management in agriculture.

45. How can ethical dilemmas related to genetically modified crops be addressed in disease resistance?

Ethical dilemmas related to genetically modified (GM) crops and disease resistance can be addressed through

1. **Transparency** Openly communicate the benefits and risks of GM crops, emphasizing disease resistance advantages.
2. **Independent Research** Support independent, long-term research on GM crop safety and environmental impacts.

3. **Inclusivity** Involve diverse stakeholders in decision-making, including farmers, scientists, policymakers, and consumers.
4. **Regulation** Implement stringent regulations to ensure the safety and ethical use of GM crops.
5. **Education** Promote public and farmer education on GM crop benefits, ethical considerations, and coexistence with non-GM crops.
6. **Choice** Respect consumers' right to choose between GM and non-GM products through labeling and market segregation.

Balancing ethical concerns with the potential benefits of GM crops helps navigate the complex landscape of disease resistance in agriculture.

46. In what ways does data privacy and security impact disease forecasting and pathometry practices?

Data privacy and security are crucial in disease forecasting and pathometry

1. **Data Protection** Ensures sensitive disease-related data, such as crop health records and remote sensing information, are safeguarded against unauthorized access or misuse.
2. **Collaboration** Secure data sharing mechanisms enable researchers and stakeholders to collaborate effectively while protecting proprietary information.
3. **Public Trust** Maintaining privacy instills trust in data collection efforts, encouraging farmers and communities to participate in disease reporting.
4. **Ethical Considerations** Ensures that data handling aligns with ethical standards and regulations, respecting individuals' privacy rights.

Balancing data privacy and security with the need for collaborative research is essential for the successful, responsible management of plant diseases in agriculture.

47. How can international collaboration enhance disease monitoring and management efforts worldwide?

International collaboration can significantly enhance disease monitoring and management efforts worldwide through

1. **Data Sharing** Facilitating the exchange of disease data, research findings, and best practices among countries for a broader understanding of global disease dynamics.
2. **Resource Pooling** Sharing resources, such as funding, expertise, and technologies, allows for more comprehensive disease research and control.
3. **Early Warning Systems** Establishing global early warning systems that provide timely alerts about emerging diseases, enabling proactive measures.
4. **Research Networks** Collaborative research networks promote the development of innovative disease management strategies.
5. **Capacity Building** Supporting capacity-building programs in developing regions to strengthen disease management infrastructure.

Such collaboration fosters a collective response to global disease threats, ensuring food security and sustainable agriculture.

Crop-Specific Considerations

48. What are the specific disease challenges faced by crops grown in greenhouse environments, and how are they managed?

Crops grown in greenhouses face specific disease challenges, including

1. **Humidity** High humidity promotes fungal diseases like powdery mildew and grey molds.
2. **Temperature** Variations in temperature can stress plants and make them susceptible to diseases.
3. **Confinement** Close proximity of plants in a greenhouse can facilitate rapid disease spread.

Management strategies include

1. **Ventilation** Proper airflow reduces humidity and minimizes disease pressure.
2. **Sanitation** Regular cleaning and removal of infected plant material prevent pathogen buildup.
3. **Resistant Varieties** Planting disease-resistant cultivars reduces the risk of infection.
4. **Bio control** Beneficial insects and microorganisms can be used to control greenhouse pests and diseases, minimizing chemical use.
5. **Integrated Pest Management (IPM)** A holistic approach combining cultural, biological, and chemical methods tailored to greenhouse conditions.

49. How do organic farming practices impact the prevalence and management of soil-borne diseases?

Organic farming practices influence soil-borne disease prevalence and management

1. **Reduced Prevalence** Organic methods, including crop rotation and cover cropping, disrupt disease cycles and reduce the build-up of soil-borne pathogens.
2. **Enhanced Soil Health** Organic practices improve soil structure and beneficial microorganism populations, making it less hospitable to pathogens.
3. **Bio control** Organic farming often employs beneficial microorganisms and nematodes to suppress soil-borne diseases.
4. **Chemical-Free** Organic farming avoids synthetic soil fumigants and chemical treatments that may harm soil health and exacerbate disease issues.

Overall, organic farming promotes healthier soils and reduces the prevalence of soil-borne diseases, emphasizing prevention and sustainability in disease management.

50. Can you explain the concept of "resistance breeding" and its significance in disease management?

Resistance breeding, also known as host resistance, is a disease management strategy in agriculture that involves developing and cultivating crop varieties that are genetically resistant to specific diseases. This approach is significant in disease management for several reasons

1. **Reduced Chemical Dependency** Resistant varieties decrease the need for chemical pesticides, contributing to environmentally friendly and sustainable farming practices.
2. **Long-Term Sustainability** Resistance traits can persist over generations, providing long-term disease protection and ensuring crop viability.
3. **Cost-Effective** Once developed, resistant varieties can be cost-effective as they require fewer inputs for disease control.
4. **Minimized Yield Loss** Resistant crops experience fewer disease-related losses, leading to increased yields and food security.
5. **Enhanced Crop Health** Disease-resistant varieties promote overall crop health and reduce the risk of epidemics.

Resistance breeding is a key component of integrated pest management (IPM) and sustainable agriculture, offering effective and eco-friendly disease management solutions.

51. What are the disease considerations for aquaponics systems, and how are they managed?

Disease considerations for aquaponics systems include

1. **Waterborne Pathogens** Aquaponics' recirculating water can harbor pathogens like bacteria and fungi.
2. **Fish Health** Diseases in fish can affect plant health through nutrient supply.
3. **Plant Susceptibility** Certain crops may be more vulnerable to waterborne diseases.

Management strategies involve

1. Biosecurity Implementing strict biosecurity measures to prevent pathogen introduction.
2. Water Quality Maintaining optimal water quality, including pH and temperature, to discourage pathogen growth.
3. Fish Health Monitoring Regularly monitoring fish health to prevent disease outbreaks.
4. Quarantine Isolating new fish to prevent potential pathogen transmission.
5. Hygiene Ensuring clean, sanitized equipment and systems to minimize disease risks in aquaponics.

52. How do advancements in genetic editing technologies, like CRISPR-Cas9, impact disease resistance research and crop improvement?

Advancements in genetic editing technologies like CRISPR-Cas9 have revolutionized disease resistance research and crop improvement

1. **Precision Editing** CRISPR enables precise modification of genes related to disease resistance, allowing breeders to create crops with enhanced immunity.
2. **Speed and Efficiency** Genetic editing accelerates the development of disease-resistant crop varieties, reducing breeding timelines.
3. **Targeted Resistance** Specific resistance genes can be introduced or modified, providing tailored protection against known pathogens.

4. **Reduced Chemical Use** Disease-resistant crops reduce reliance on chemical pesticides, promoting eco-friendly agriculture.
5. **Crop Diversity** Genetic editing expands the potential for disease-resistant varieties across diverse crops, ensuring global food security.

These technologies empower researchers and breeders to address disease challenges efficiently, fostering more resilient and productive agriculture.

53.What are the potato disease forecasting models used in India

- **JHULSACAST Model**: This model was originally developed for western Uttar Pradesh and later modified and adapted for other regions like West Bengal and Punjab.
- **INDO-BLIGHTCAST Model**: This is a web-based forecasting model developed by the All India Coordinated Research Project (AICRP) on Potato. It uses daily weather data from meteorological stations across India to predict the first appearance of late blight. It is a pan-India model, meaning it doesn't require local calibration and can be used in different regions.

54. What are other plant disease forecasting models used in India

- Rice Blast Advisory Model (RBAM)
- BlastSIM Model
- Brown Rust Warning System
- Groundnut Rust Early Warning System
- Downy Mildew Forecasting in Pearl Millet

55. What were Mills rule for apple scab

Mills' rules, developed in 1944 by American plant pathologist W.D. Mills, are a set of guidelines used to predict apple scab infection periods based on temperature and leaf wetness duration. The table correlates average temperature during leaf wetness with the minimum number of hours of continuous wetness required for apple scab infection to occur. The rules classify infection risk into three categories:

- **Low**: Minimal infection is likely to occur.
- **Moderate:** Some infection may occur, and preventive measures should be considered.
- **High:** Significant infection is likely, and fungicide application is recommended

8

Genome Sequencing, Phylogenetic Analysis, Intellectual Property Right (IPR)

1. What is Genome Sequencing?

Genome sequencing is the process of determining the complete DNA sequence of an organism's genome. The genome is the genetic material that contains all the instructions necessary for the growth, development, functioning and reproduction of an organism. It is composed of DNA (deoxyribonucleic acid., which is a long molecule made up of four building blocks called nucleotides adenine (A), thymine (T), cytosine (C), and guanine (G).

2. Why is genome sequencing important in plant pathology?

Genome sequencing is essential in plant pathology for several reasons

i. Genome sequencing in plant pathology identifies and characterizes plant pathogens at the molecular level.
ii. It helps trace the evolution of pathogens and their genetic adaptations.
iii. Genomic data enables the development of precise diagnostic tools for rapid pathogen detection.
iv. Understanding host-pathogen interactions informs disease management strategies.
v. Genomic information aids breeding disease resistance crop varieties.
vi. It monitors pathogen diversity and population dynamics
vii. Genomic data supports global surveillance and response to emerging threats.
viii. Biosecurity and quarantine measures can be guided by genomic information to protect agriculture.
ix. Genome sequencing contributes to environmentally friendly disease management approaches.
x. It enhances our knowledge of plant diseases and their impacts on crops and ecosystems.

3. Historical milestones of genome sequencing in context of plant pathology?

Ans. In context of plant pathology, there have been several key historical milestones related to genome sequencing. These milestones have significantly advanced our understanding of plant-pathogen interactions and have implications for agriculture and crop protection. Here are some historical milestones specific to plant pathology

1. 1982 The genome of *Agrobacterium tumefaciens*, a bacterium causing crown gall diseases in plants, was sequenced. This marked one of the first instances of genome sequencing in plant pathology.
2. 1996 Complete genome sequencing of *Xylella fastidiosa,* a fastidious bacterium causing several diseases in grapevine and citrus.
3. 2001 The genome of *Phytophthora infestans*, the oomycete responsible for the potato Irish famine and late blight disease, was sequenced. This insights into genetic basis of the pathogen's virulence and host interactions.
4. 2006 The genome of *Magnaporthe oryzae*, the fungus causing rice blast, was sequenced. This pathogen poses significant threat to rice crops, and genome sequencing helped identify key pathogenicity genes.
5. 2015 The genome of *Hemielia vastatrix*, the coffee rust fungus, was sequenced. This research helped in developing resistant coffee varieties.
6. 2016 The genome of *Fusarium oxysporum* f. sp. *cubense* tropical race 4 (TR4) strain, causing Panama disease in banana plants, was sequenced. This aimed to develop resistant banana cultivars.

These milestones demonstrate the significance of genome sequencing in plant pathology, enabling researchers to identify pathogens, understand their virulence mechanisms, and develop strategies for crop protection and disease management. The genomic insights gained from these studies have direct implications for agriculture, food security, and sustainable farming practices.

4. Type of Genome sequencing

In broad sense, genome sequencing can be categorized into two main types

i. **Whole genome Sequencing (WGS)** involves the comprehensive sequencing of an organism's entire genome. It aims to decode the complete genetic information present in an organism's DNA, including coding and non-coding regions. It is used for various purpose, such as species identification, comparative genomics, and understanding genetic variations.

ii. **Targeted Sequencing** focuses on specific region of genome rather than sequencing the entire genome. It involves various approaches like

- **Exome sequencing** Selectively sequences the protein-coding region (exons) of the genome. It is commonly used for identifying genetic variants associated with diseases.
- **Amplicon sequencing:** It sequences specific DNA fragments or regions of interest. It is commonly used in studies involving targeted genes or loci.
- **Chromosomal or locus-specific Sequencing:** In this approach, specific chromosomes or genomic regions are targeted for sequencing, allowing for in-depth analysis of particular genetic elements.

5. What is conventional and modern genome sequencing?

Conventional genome sequencing method

Sanger Sequencing (Chain Termination Sequencing) was the first widely used DNA sequencing method, developed by Fedrick Sanger in the 1970s. It involves the use of modified nucleotides (di-deoxynucleotides) that terminate DNA strand elongation when incorporated into growing DNA strand. This results in a series of fragments of different lengths that are separated by gel electrophoresis and used to deduce the DNA sequence. While accurate, Sanger sequencing is relatively slow and expensive, limited its use for large scale genome projects.

Modern Genome Sequencing

1. **Next generation Sequencing (NGS)** NGS technologies, which emerged in the 2000s, have revolutionized genome sequencing. These methods are characterized by their high-throughput, parallel processing, and cost-effectiveness. Key NGS technologies include
   - **Illumina Sequencing** This is one of the most widely used NGS platforms. It employs reversible terminators and sequencing by synthesis chemistry. It can generate massive amount of sequencing data in a short tie, making them suitable for a wide range of applications.
   - **PacBio Sequencing** Pacific Biosciences (PacBio) sequencing is based on single-molecule, real time (SMRT) sequencing technology. It offers longer read lengths compared to Illumina, which is beneficial for de novo genome assembly and structural variant analysis.
   - **Oxford Nanopore Sequencing** This technology uses nanopores to directly sequence DNA molecules. It can produce long reads and is portable, making it suitable for fieldwork and rapid sequencing.
2. **Third-Generation Sequencing** Beyond NGS, there is an ongoing development of third generation sequencing technologies, which aim to produce even longer read lengths and reduce sequencing errors. These technologies, such as the improved versions of PacBio and nanopore sequencing, are being adopted for more comprehensive genome assemblies and structural variant detection.
3. **Metagenomic Sequencing** This approach involves sequencing entire communities of microorganisms from environmental samples. It has applications in studying microbial diversity and ecology, such as soil, water microbiome.
4. **Single cell sequencing** It helps to sequence the DNA of individual cell, providing insights into cellular heterogeneity within tissues or organisms.

6. What is the procedure of genome sequencing?

Overview of the genome sequencing process

1. **DNA Extraction:** The first step involves isolating the DNA from the organism's cells. This may involve various methods and quality control measures to ensure the purity and integrity of the DNA.
2. **DNA fragmentation:** The extracted DNA is prepared for sequencing. This involves fragmentation of the DNA into smaller, manageable fragments. Adapters with unique barcodes are often ligated to the ends of these fragments to enable sample identification in multiplexed sequencing.
3. **Sequencing Process:** The DNA fragments are amplified and sequenced using chosen technology. Modern sequencing machines can generate massive amounts of sequencing data quickly and cost effectively.

4. **Data Analysis:** The raw sequencing data is processed and analyzed using specialized software and algorithms to assemble the short sequencing reads into complete genome sequence. This involves aligning and comparing the reads to a reference genome.
5. **Annotation:** Once the genome sequence is obtained, it is annotated to identify genes, regulatory elements, and other functional elements within the genome. This step helps to understand the biological significance of different regions of the genome.

7. What is the use of DNA sequencing?

By the use of DNA sequencing we can predict transmembrane domain, receptor recognition site, DNA binding region, regulation of gene.

8 Why oligonucleotide primer is used in dideoxynucleotide procedure?

It provides a 3' hydroxyl group for the initiation of DNA synthesis.

9. In Sanger's method of DNA sequencing chain, which is used for termination?

In Sanger's method of DNA sequencing chain termination is achieved by using ddNTPS

10. What is the basis of pyrosequencing technique?

The basis of pyrosequencing technique is detection of pyrophosphate

11. The pyrophosphate released during pyrosequencing are

Beta and γ phosphate of incoming nucleotide

12. Name a cellular organism was whose genome was sequenced?

Haemophilus influenza

13. Frederick Sanger and Allan Maxam and Walter Gilbert developed DNA sequencing on which organism?

Complete DNA genome sequencing of bacteriophage ϕX 174 (1977)

14. Name of first plant pathogen genome sequenced?

Agrobacterium tumefaciens was one of the first plant pathogen to have its genome sequenced in 1982. It is a bacterium responsible for crown gall disease in plants.

15. Name the first plant pathogenic bacteria whose genome has been completely sequenced?

In 1996, in *Xylella fastidiosa* was first plant pathogenic bacteria whose complete genome sequencing was done.

16. Name of first plant pathogenic fungus to have its genome sequenced?

Magnaporthe oryzae (Rice blast)

17. Name of the first fungus whose genome was completely sequenced?

The first fungal genome to be sequenced was that of *Saccharomyces cerevisiae,* commonly known as baker's yeast. This milestone occurred in 1996, making it one of the earliest complete genome sequences of any organism.

18. What is Bio informatics?

Bio informatics is the field of science in which biology, computer science and information technology merge into a single discipline (NCBI 2001). It develops software and algorithms. It is used to gather, analyze and merge biological information. It used in many fields like crop improvement, biotechnology, evolutionary studies etc.

19. What is application of Bio informatics ?

Bio informatics in genome sequencing involves using computational tools and methods to process, analyze, and interpret genomic data. It plays a crucial role in tasks like sequence assembly, quality control, variant calling, gene annotation, comparative genomics, and more, facilitating the extraction of meaningful insights from genomic information.

20. Enlist the Bio informatics resources in India?

In India, prominent Bio informatics include institutions like the National Centre for Biological Sciences (NCBS) and the Centre for Cellular and Molecular Platforms (C-CAMP), as well as websites like Bio informatics National Certification (BINC. Examination and the Indian Bio informatics Consortium (INBIC.. These resources offer a range of services, research opportunities, and educational programs in field of Bio informatics .

21. What is the role of Bio informatics in genome sequencing?

It plays an important role in genome sequencing

1. Manages and organizes massive sequencing data.
2. Assembles short DNA sequences into complete genomes.
3. Corrects errors in sequencing data.
4. Identifies genetic variations {SNPs, insertions-deletions (indels)} .
5. Annotates gene functions and regulatory elements.
6. Compares genomes for evolutionary insights.
7. Analyses complex environmental samples (metagenomics).
8. Constructs phylogenetic trees to depict evolutionary relationships.
9. Predicts protein and RNA structures.
10. Integrates genomic data with other omics for functional insights.

22. What are the nucleotide and protein databases?

Nucleotide databases are European molecular biology laboratory (EMBL), gene bank, DNA Data bank of Japan (DDBJ) and protein databases are sequence UNIPORT, PIR, SWISS PORT, Interaction biological string, Structure PDB- Protein data base, CATH-online resource, SCOP- Structural classification of protein.

23. What is sequence assembly?

Sequence assembly is the process of reconstructing complete genetic sequences from short DNA or RNA fragments. It involves aligning and merging these fragments to create longer sequences, known as contigs or scaffolds. This is essential for tasks like genome sequencing and transcriptome analysis. Two types of assembling, *De novo* assembly assembles short reads to create full length assembly and mapping assembly reads against existing gene sequence. Software used for assembly are Bio-edit, MEGA 4, Gene construction kit, Peaks, Arachne.

24. What is sequence alignment?

Sequence alignment is a fundamental technique in Bio informatics and computational biology used to compare and analyze biological sequences, such as protein, RNA and DNA sequences. Its aim is to identify similarities and differences between sequences, which can provide insights into their evolutionary relationships, functional similarities, and structural properties.

25. What are the methods used for alignment of sequences?

These are the common methods used for sequence alignment

1. Pairwise Sequence alignment

Needelman-Wunch algorithm performs global sequence alignment, ensuring that the entire length of both sequences is considered in the alignment.

Smith waterman algorithm is used for local sequence alignment, identifying regions of similarity within sequences.

2. Multiple Sequence Alignment (MSA.

Clustal W and Clustal Omega These tools perform multiple sequence alignments of three or more sequences, considering both pairwise and progressive alignment methods.

MUSCLE (multiple Sequence Comparison by Log-Expectation) MUSCLE is another popular tool for MSA.

MAFFT (Multiple Alignment using Fast Fourier Transform) MAFFT employs a variety of alignment strategies, including progressive alignment.

3. Profile based Alignment

BSI-BLAST (Position-Specific Iterated BLAST) PSI-BLAST builds a position-specific scoring matrix (PSSM) from a set of related sequences and uses it to improve sensitivity in databases searches.

4. Local Alignment Some methods, like BLAST (Basic Local Alignment Search Tool), are optimized for finding local similarities between sequences within a large database.

5. Heuristic methods Some alignment tools use heuristic approaches to reduce computation time while still providing reasonably accurate alignments, such as FASTA.

26. What is genome annotation?

Genome annotation is the process of identifying the locations of genes and all of the coding regions in the genome and understanding what those genes do. This can be done by 2 two types, structural annotation and functional annotation.

27. Name the software used for genome annotation?

Plants genome annotation software is Genome Threader, for eukaryotes and prokaryotes is Manatee, CRITICA and only for eukaryotes genome annotation is Glimmer HMM, Genzilla, JIGsaw.

28. What do you understand by Phylogenetic analysis?

Phylogenetic analysis, also known as phylogenetics, is a field of biology that focuses on studying relationships and history of species, genes, or other biological entities. It aims to construct phylogenetic trees or networks that depict the evolutionary branching patterns and common ancestors of these entities. It is fundamental in understanding evolutionary processes, genetic diversity, and biological classification.

29. What is phylogeny?

Phylogeny is the study of the evolutionary history and relationships among different species or groups of organisms. This field uses methods like genetic analysis to create phylogenetic trees, diagrams that show the branching patterns of evolutionary relationships. It's important for understanding biodiversity, evolution, and related biological sciences.

30. Enlist the tools used for phylogenetics?

BioNumerics, Dendroscope, MEGA, Morpho Bank, Tree finder, Darwin

31. How many types of interference are in phylogeny?

There are two types of interference rooted phylogeny and unrooted phylogeny

Rooted phylogeny in which the phylogenetic tree contains a basal point or common ancestor. It contains terminal and internal points whereas unrooted phylogeny It does not contain a basal point. It does not contain terminal and internal points.

32. What is clustering and its types?

Clustering is task of grouping a set of objects and each set is named by cluster. It helps to analyze the large amount of data in simple way. Four types of clustering are Hierarchical, Centroid, Distribution, Density.

33. What do you understand by Intellectual Property Rights (IPR)?

According to World intellectual property organization (WIPO), Intellectual property right (IPR) refers to legal rights that protect creations of the mind, such as inventions, artistic works, and symbols. IPRs are granted to individuals, organizations, and businesses.

IPR refers to legal ownership of intellectual property by person or business of an invention, discovery related to the particular product or processes for the protection of the owner against unauthorized copying or imitations.

34. When did first IPR came into force in India?

In 1856, first India Patent Act, was introduced which remained in force for more than 50 years which was later modified and revised and was called "The Indian Patents and Designs Act, 1911". A complete bill on patent rights was enacted after Independence in the year 1970 and was called "The Patents Act, 1970".

35. What is aim of The Patent Act, 1970?

The Indian Patent Act, 1970, aims to encourage innovation and economic growth by providing inventors with exclusive rights to their inventions for a limited time. It also promotes knowledge sharing and balances inventors' rights with the public interest. The act aligns with international patent standards and includes provisions to protect traditional knowledge and provide exceptions for public health concerns, making it a crucial component of India's intellectual property framework.

36. How many types of Intellectual property?

Mainly two broad categories are industrial property and non-industrial property. Under industrial property is patent, design, trademark, geographical indications whereas copyright comes under non- industrial property.

37. Define patent?

A patent is the exclusive right granted by a government authority to an inventor to manufacture, use, or sell an invention and prohibit others from making, using or selling an invention for a limited period of time.

38. Define designs?

It is art or practice of designing object for manufacturing consisting aesthetic aspects and three-dimensional features like shape, surface, patterns, lines, colors or two dimension features of objects.

39. Define trademark?

It is a symbol, word, or words registered legally or established by use as to represent a company or product.

40. What do you understand by Geographical indication?

It is a name or sign used on products which corresponds to a specific geographical location or origin having specific qualities, reputation or characteristics.

41. Define copyright?

It is a legal provided to creators to perform, print, publish, record literacy, artistic, or musical material for its uses and distributions.

42. What is the use of trademark?

Trademark is represented graphically. It is capable of distinguishing the goods or services of one person from those of others. It may include shapes of goods or combination of colors.

43. When did Copyright Act, came into force?

Copyright Act, 1957

44. What do you understand by The Agreement on Trade-Related Aspects of Intellectual Property Rights (TRIPS) ?

TRIPS came into effect in 1 January, 1995. It is an international agreement administered by the World Trade Organization (WTO) that sets down minimum standards for many forms of intellectual property (IP) regulations for all the members of WTO.

The TRIPS agreement introduced intellectual property law into the international trading system for the first time and remains the most comprehensive international agreement on intellectual property in India.

45. Give examples of geographical indication in India.

Darjeeling tea, Nagpur orange, Kanchipuram Silk, Tripura Queen Pineapple, Navara Rice, Alphonso mango, Basmati Rice, Kinnauri Shawl.

46. What is Plant Variety Protection?

Plant Variety Protection (PVP) also called Plant Breeders' Rights (PBR), started in 1987. This is a legal framework that grants exclusive rights to plant breeders for a specified period (typically 20-25 years) to control the production, sale and distribution of a novel plant variety they have developed. It requires variety to be new, distinct, uniform, and stable and it varies by country with exemptions for farmers. It is distinct from patents and aims to protect plant varieties rather than inventions or discoveries.

47. What is Protection of Plant Rights and Farmers Rights (PPV&FR)?

Aim of Protection of Plant Rights and Farmers Rights (PPV&FR) Act, 2001 is to recognize and protect the rights of farmers in respect of their contributions made at any time in conserving, improving and making available plant genetic resources for the development of new plant varieties.

48. Give example of copyright?

Copyright gives protection to both tangible forms, such as written manuscripts or physical artworks, as well as intangible digital content like online articles, images, videos, and software applications.

49. Example of trademark in agriculture?

Trademark eg. "Cherry delight" for Organic Cherries. Its aim is to create a unique and recognizable brand for their organic cherries.

9

Plant Defense Mechanism

- Plants have structural and biochemical defense mechanisms against pathogens.
- Structural defense mechanisms can be pre-existing (e.g. wax, thick cuticle) or post-infectional (e.g. cork layer formation, tyloses).
- Pre-existing biochemical defenses include inhibitors and phenolics; post-infectional defenses include phytoalexins and plantibodies.
- Disease development depends on a successful host-pathogen interaction; defense mechanisms are ineffective during a compatible reaction.

Plant Defense Mechanisms

Defense Type	Mechanism	Examples
Physical Defenses	Structural features that deter or impede threats	Thorns, spines, trichomes, waxy cuticle, thick bark, silica deposits
Chemical Defenses	Production of toxic or repellent compounds	Alkaloids (caffeine, nicotine), terpenoids (essential oils), phenolics (tannins), phytoalexins, cyanogenic glycosides
Induced Defenses	Activation of specific responses to threats	Jasmonic acid, salicylic acid, systemic acquired resistance (SAR), hypersensitive response (HR)
Indirect Defenses	Attraction or harboring of beneficial organisms	Volatile organic compounds (VOCs), extrafloral nectaries, mutualistic relationships with fungi or bacteria

1. What is the role of a plant's surface in defense against pathogens?

The plant's surface acts as the first line of defense, as pathogens must adhere to and penetrate this barrier to cause infection.

2. What are the two main types of structural defense mechanisms in plants?

Pre-existing and post-infection or induced structural defense mechanisms.

3. Give examples of pre-existing structural defence

Wax, thick cuticle, thickness and toughness of the outer wall of epidermal cells, stomata, sclerenchyma cells, and lenticels.

4. How does wax contribute to plant defense?

Wax forms a protective coating on plant leaves and fruit, which is synthesized by the epidermis and is extremely hydrophobic, thereby repelling water and pathogens.

5. How do the cuticle and epidermal cells contribute to plant defense?

The thickness and toughness of the cuticle and epidermal cell walls can hinder pathogen penetration. For example, resistance to *Puccinia graminis tritici* in Barberry species is attributed to tough outer epidermal cells with a thick cuticle.

6. What role do sclerenchyma cells play in plant defense?

Sclerenchyma cells, with their thickened walls of lignin, provide mechanical support and effectively block the spread of some fungal and bacterial pathogens.

7. How do stomata influence plant defense?

Stomata are natural openings through which many pathogens enter plants. The structure of stomata can provide resistance to penetration by certain pathogens.

8. What are lenticels, and how do they relate to plant defense?

Lenticels are openings on fruit, stem, and tubers that allow air passage. Their shape and internal structure can influence the incidence of fruit diseases.

9. What triggers post-infectional or induced structural defense mechanisms?

Pathogen penetration through the host surface induces structural defense mechanisms in the host cells.

10. What are the two categories of post-infectional or induced structural defenses?

Histological defense barriers (e.g., cork layer, abscission layers, tyloses formation) and cellular defense structures (e.g., hyphal sheathing).

11. How does the formation of a cork layer protect plants?

Infection by pathogens can induce plants to form cork cells beyond the infection point, inhibiting further pathogen invasion and blocking the spread of toxins.

12. What is the purpose of abscission layers in plant defense?

Abscission layers are gaps formed between infected and healthy cells, causing the infected area to shrivel, die, and slough off, carrying the pathogen with it.

13. How do tyloses contribute to plant defense?

Tyloses are overgrowths of protoplast from adjacent living cells that protrude into xylem vessels, blocking the further advance of pathogens.

14. What is the role of gum deposition in plant defense?

Ans. Gums are produced by many plants around lesions after infection or injury, often under stress conditions. They can act as a physical barrier against pathogens.

15. What is Hyphal sheathing?

Hyphal sheathing is a cellular defense structure observed in plants like flax infected with *Fusarium oxysporum* f.sp. *lini*, where the plant cells form a sheath around the invading hyphae.

16. What are the two types of pre-existing biochemical defenses in plants?

Inhibitors released by the plant into its environment and inhibitors present in plant cells before infection.

17. How do inhibitors released by plants protect against pathogens?

These substances can directly affect microorganisms or encourage the growth of beneficial microbes that act as antagonists to pathogens.

18. Give examples of inhibitors present in plant cells before infection.

Phenolics, tannins, glucanases, dienes, and chitinase.

19. How do phenolic compounds contribute to plant defense?

Phenolic compounds, such as catechol and protocatechuic acid in onions, can inhibit the growth and development of pathogens.

20. What are saponins, and how do they function in plant defense?

Saponins are antifungal compounds with membranolytic activity, disrupting the cell membranes of fungi. Examples include tomatine in tomato and avenacin in oats.

21. What are lectins, and what is their role in plant defense?

Lectins are proteins that bind to specific sugars and occur in high concentrations in many seeds. They can cause lysis and growth inhibition of many fungi.

22. How do hydrolytic enzymes contribute to plant defense?

Enzymes like glucanases and chitinases can break down the cell walls of pathogens, inhibiting their growth and development.

23. What are some specific examples of pre-existing structural defense mechanisms in plants?

Here are some specific examples of pre-existing biochemical defense mechanisms in plants

- Inhibitors released by the plant Substances like α-terthinyl from marigold roots, which is inhibitory to nematodes, and maleic acid from glandular hairs of chickpea, which inhibits spore germination.
- Inhibitors present within plant cells
- Phenolic compounds like catechol and protocatechuic acid in onions.
- Chlorogenic acid in potato.
- Saponins like tomatine in tomato and avenacin in oats.
- Lectins, which are proteins that bind to specific sugars.
- Hydrolytic enzymes like glucanases and chitinases.

24. What are some specific examples of pre-existing structural defense mechanisms in plants?

Here are some specific examples of pre-existing structural defense mechanisms in plants

- **Thick Cuticle** The thickness and toughness of the cuticle, a waxy layer covering the epidermis, can hinder pathogen penetration. For example, resistance to *Puccinia graminis tritici* in Barberry species is attributed to tough outer epidermal cells with a thick cuticle.
- **Cell Wall Composition** The composition and structure of plant cell walls, including the presence of lignin and silica, can provide resistance to pathogen penetration. For example, silicification and lignification of epidermal cells offer protection against *Pyricularia oryzae* and *Streptomyces scabies* in paddy and potato, respectively.

- **Stomata** The shape and structure of stomata, which are natural openings on plant surfaces, can influence the ability of pathogens to enter the plant. For example, the citrus variety szinkum is resistant to citrus canker due to the structure of its stomata.
- **Lenticels** The shape and internal structure of lenticels, which are openings on fruit, stems, and tubers, can affect the incidence of fruit diseases. For instance, small and suberized lenticels offer resistance to the potato scab pathogen *Streptomyces scabies*.

25. What are some specific examples of Post-infection structural defense mechanisms in plants?

- Post-infection structural defenses are the plant's defense responses that are activated after the pathogen has already established infection. Here are some examples
- **Cellular defense structures** such as hyphal sheathing, which is observed in flax infected with *Fusarium oxysporum* f.sp. *lini*.
- **Histological defense structures** such as the formation of cork layer, abscission layer, and tyloses.
- **Cork layer** Several layers of cork cells are formed beyond the point of infection to inhibit further invasion by the pathogen and block the spread of toxin substances secreted by the pathogen. It also stops the flow of nutrients and water from the healthy to the infected area.
- **Abscission layer** A gap is formed between infected and healthy cells of leaf surrounding the locus of infection due to the disintegration of the middle lamella of parenchymatous tissue. The infected area shrivels, dies, and sloughs off, carrying with it the pathogen.
- **Tyloses** Overgrowths of the protoplast of adjacent living parenchymatous cells protrude into xylem vessels through pits. They have cellulosic walls and form quickly ahead of the pathogen, clogging the xylem vessels and blocking the further advance of the pathogen in resistant varieties.
- **Gum deposition** Various types of gums are produced by many plants around lesions after infection by a pathogen or injury. Gum secretion is most common in stone fruit trees but occurs in most plants, generally under stressed conditions.

26. What are pre existing chemical defense in plants ?

Pre-existing chemical defense in plants

- Inhibitors - Released by plants into the environment, and present in plant cells before infection.
- Phenolics - Tannins, Glucanases, Dienes, Chitinase

27. What are post infection or induced biochemical defense in plants?

Post-infection or induced chemical defense in plants are

- **Hypersensitivity response (HR)** - Localized induced cell death in the host plant at the site of infection by a pathogen, thus limiting the growth of the pathogen.

- **Production of Antimicrobial substances -**
- **Phytoalexins** - Toxic antimicrobial substances produced in appreciable amounts in plants only after stimulation by phytopathogenic microorganisms or by chemical or mechanical injuries
- **Plantibodies** - Antibodies encoded by animal genes, but produced in and by plants.

27. What is induced resistance?

- It is a plant defense mechanism triggered by prior infection or treatment.

28. How long has induced resistance been studied?

- For over 100 years.

29. What are the two main types of induced resistance?

- Systemic Acquired Resistance (SAR) and Induced Systemic Resistance (ISR).

30. What is SAR?

- It is a broad-spectrum resistance that occurs following a localized infection.

31. What is ISR?

- It is triggered by specific beneficial microbes and primes the plant for faster defense responses.

32. What is the difference between SAR and ISR?

- SAR is SA-mediated, while ISR is JA and ethylene-mediated.

33. How does SAR work?

- It involves salicylic acid signaling and the expression of pathogenesis-related (PR) proteins.

34. How does ISR work?

- It primes the plant for a faster and stronger defense response upon pathogen attack.

35. What are PR proteins?

- They are antimicrobial proteins produced by plants in response to infection.

36. What are some examples of biological agents that induce resistance?

- Beneficial bacteria and fungi.

37. What is the role of signaling in induced resistance?

- Signals like salicylic acid and jasmonic acid coordinate defense responses.

38. Can induced resistance be used in crop protection?

- Yes, it is a promising strategy for sustainable agriculture.

39. What is the historical background of induced resistance?

- It was first observed in the early 1900s by Ray (1901) and Beauvene.

40. Who conducted the first controlled study of SAR?

- Ross (1961) demonstrated SAR in tobacco plants.

41. What is Local Acquired Resistance (LAR)?

- It is a localized resistance response around the infection site.

42. Does induced resistance provide protection against a broad range of pathogens?

- Yes, it can be effective against various pathogens, including viruses, bacteria, and fungi.

43. Is phytotoxicity a concern with induced resistance?

- Phytotoxicity is more associated with SAR than ISR.

44. Are all plant-pathogen interactions lead to induced resistance?

- No, only incompatible interactions typically lead to induced resistance.

45. What are some challenges in using induced resistance in agriculture?

- The effectiveness can vary depending on the plant, pathogen, and environmental conditions.

10

Plant Disease Resistance and Its Mechanisms

What is Plant Disease Resistance and Its Mechanisms

Plant disease resistance is the ability of a plant to prevent or reduce the damage caused by pathogens (disease-causing organisms) like fungi, bacteria, viruses, and nematodes. It involves a complex interplay between the plant's defense mechanisms and the pathogen's virulence factors.

There are two main types of plant disease resistance:

Innate Resistance (Non-Host Resistance): This is a broad-spectrum resistance that prevents most potential pathogens from infecting the plant. It is often due to the lack of essential factors required for the pathogen's survival or the presence of preformed physical and chemical barriers.

Induced Resistance: This type of resistance is activated in response to pathogen attack.It involves the recognition of pathogen-associated molecular patterns (PAMPs) or effectors, triggering a cascade of defense responses that can limit or stop the pathogen's growth and spread

Mechanisms of Plant Disease Resistance

Preformed Defenses

- Physical barriers: Waxy cuticle, thick cell walls, stomata closure
- Chemical barriers: Antimicrobial compounds (phytoalexins, phenolics), toxic proteins, enzyme inhibitors
- Existing proteins and enzymes: Pathogenesis-related (PR) proteins, reactive oxygen species (ROS)

Induced Defenses

- Recognition of PAMPs/effectors by plant receptors
- Signal trans duction pathways: Activation of defense-related genes
- Production of antimicrobial compounds and proteins
- Hypersensitive response (HR): Localized cell death to contain infection
- Systemic acquired resistance (SAR): Long-lasting, broad-spectrum resistance induced by a previous infection

Gene-for-Gene Resistance

- Specific recognition of pathogen effectors by plant resistance (R) genes
- Triggers a rapid and strong defense response, often including HR

Quantitative Resistance

- Multiple genes contribute to partial resistance
- Reduces pathogen growth and disease severity
- Disease Resistance is the ability of an organism to completely or partially exclude or overcome the effect of a pathogen or other damaging factor.
- **There are two types of resistance** Vertical Resistance and Horizontal Resistance.
- Vertical Resistance is specific to a particular pathotype and is governed by oligogenic genes. It is characterized by a hypersensitive response and is relatively easy to select and evaluate.
- **Horizontal Resistance** is non-specific to pathotypes and is governed by polygenic genes. It is characterized by a resistant response and is difficult to select and evaluate.
- Mechanisms of Disease Resistance include disease escape, disease endurance or tolerance, immunity, and hypersensitivity.
- Disease escape is the ability of susceptible host plants to avoid disease attack due to environmental conditions, early varieties, changes in planting dates or sites, and balanced application of NPK fertilizers.
- Disease endurance or tolerance is the ability of plants to tolerate pathogen invasion without showing much damage.
- Immunity occurs when the host does not show disease symptoms, often due to the prevention of the pathogen from reaching the appropriate host parts.
- Hypersensitivity is a resistant response where host cells surrounding the infection point die, leading to the death of the pathogen.
- Phytoalexins are specific chemicals produced by the host in response to pathogen infection.
- Sources of Disease Resistance include germplasm collections, related species, mutations, and known varieties.

1. What is Disease Resistance?

- The ability of an organism to withstand or overcome the effects of a pathogen.

2. What are the two types of Disease Resistance?

- Vertical Resistance and Horizontal Resistance.

3. What is Vertical Resistance?

- Resistance specific to a particular pathotype.

4. What is Horizontal Resistance?

- Resistance that is not specific to pathotypes.

5. What genes govern Vertical Resistance?

- Oligogenic genes.

6. What genes govern Horizontal Resistance?

- Polygenic genes.

7. How is Vertical Resistance characterized?

- By a hypersensitive response.

8. How is Horizontal Resistance characterized?

- By a resistant response.

9. Is Vertical Resistance easy or difficult to select and evaluate?

- Relatively easy.

10. Is Horizontal Resistance easy or difficult to select and evaluate?

- Difficult.

11. What are the mechanisms of Disease Resistance?

- Disease escape, disease endurance or tolerance, immunity, and hypersensitivity.

12. What is Disease Escape?

- The ability of susceptible plants to avoid disease due to environmental factors or farming practices.

13. What is Disease Endurance or Tolerance?

- The ability of plants to withstand pathogen invasion without significant damage.

14. What is Immunity?

- When the host shows no disease symptoms.

15. What is Hypersensitivity?

- A resistance response where host cells around the infection die, killing the pathogen.

16. What are Phytoalexins?

- Specific chemicals produced by the host in response to infection.

17. What are the sources of Disease Resistance?

- Germplasm collections, related species, mutations, and known varieties.

18. Where can a breeder find needed resistance if it's already documented?

- In a known variety.

19. Where should one look for resistance to a new disease or pathotype?

- Germplasm collections.

20. How can resistance be transferred from related species?

- Through interspecific hybridization.

21. Can resistance be obtained through mutation?

- Yes, either spontaneously or induced.

22. What are some examples of induced resistance through mutation?

- Resistance to Victoria blight in oats, stripe rust in wheat, brown rust in oats, and mildew in barley.

23. Which rice varieties are resistant to Blast disease?

- Co25, Co26.

24. Which wheat variety is resistant to all three rusts?

- NP 809.

25. Which wheat varieties are resistant to Yellow rust?

- NP 785, NM86.

26. Which sugarcane varieties are resistant to Red rot?

- Co 419, Co 421, Co 527.

27. Which cotton varieties are resistant to Wilt?

- Vijay, Kalyan, Suyog.

28. Which cotton variety is resistant to Tikka leafspot?

- Ah 45.

29. Is Vertical Resistance specific or non-specific to pathotypes?

- Specific.

30. Is Horizontal Resistance qualitative or quantitative in phenotypic expression?

- Quantitative.

31. What is the difference between disease escape and disease endurance?

- Disease escape Susceptible plants avoid disease due to factors like environmental conditions, early maturation, or changes in planting practices. It's about avoiding the disease altogether.
- Disease endurance (or tolerance) Plants withstand the pathogen's invasion without showing significant damage. It's about the plant's ability to tolerate the disease and continue growing despite infection.

32. What factors of disease escape help in disease management?

- **Environmental conditions** This could include factors like temperature, humidity, and rainfall that make it difficult for a disease to establish itself or spread.
- **Early varieties** Planting crop varieties that mature early can help them avoid disease pressure that may come later in the growing season.
- **Changes in planting dates** Adjusting planting dates can help crops avoid periods when disease pressure is highest.
- **Changes in the site of planting** Moving crops to different locations can help avoid areas where disease inoculum is present or where environmental conditions favor disease development.
- **Balanced application of NPK fertilizers** Providing crops with proper nutrition can help them grow strong and healthy, making them less susceptible to disease

11

Terminology in Plant Pathology

1. **Pathogen:** An entity, usually a microorganism, that can cause disease in a plant.
2. **Disease:** It is a harmful alteration of the normal physiological and biochemical development of a plant (National Academy of Science, 1976).
3. **Symptom:** The external or internal reactions or alterations of a plant as a result of a disease.
4. **Sign:** The pathogen or its parts or products seen on a host plant.
5. **Syndrome:** The set of varying symptoms characterizing a disease.
6. **Inoculum:** The part of the pathogen which on contact with a susceptible host plant causes infection.
7. **Infection:** The establishment of a parasitic relationship between a pathogen and its host plant.
8. **Incubation Period:** The time between infection and the appearance of the first symptoms.
9. **Lesion:** A localized area of diseased tissue.
10. **Blight:** The rapid and extensive death of plant tissue.
11. **Canker:** A dead area on a stem, often sunken.
12. **Gall:** An abnormal swelling or outgrowth of plant tissue.
13. **Wilt** Loss of rigidity and drooping of plant parts due to inadequate water supply.
14. **Rot:** The softening and decomposition of plant tissue.
15. **Chlorosis:** Yellowing of plant tissue due to lack of chlorophyll.
16. **Necrosis:** Death of plant cells or tissues.
17. **Etiolation:** Yellowing of plant tissue due to lack of light.
18. **Stunting:** Reduced growth of a plant.
19. **Dieback:** Progressive death of shoots, branches, or roots starting at the tips.
20. **Damping-off:** Decay of seeds or seedlings before or after emergence.
21. **Sporulation:** The production of spores.
22. **Host:** A plant that is susceptible to infection by a pathogen.
23. **Resistance:** The ability of a plant to prevent or restrict the development of a disease.
24. **Susceptibility:** The inability of a plant to prevent the development of a disease.
25. **Virulence:** The degree of pathogenicity of a pathogen.
26. **Vector:** An organism that transmits a pathogen.
27. **Epidemic:** A widespread occurrence of a disease in a plant population.

28. **Endemic:** A disease that is constantly present in a moderate to severe form in a particular area.
29. **Pandemic:** A disease that is prevalent throughout a country, continent, or the world.
30. **Sporadic:** A disease that occurs at irregular intervals and locations.
31. **Hypha (plural Hyphae):** The fundamental, thread-like structural unit of a fungus.
32. **Mycelium:** A collective mass or network of hyphae, forming the main vegetative body of a fungus.
33. **Spore:** The reproductive unit of a fungus, capable of developing into a new individual.
34. **Conidium (plural Conidia:** A type of asexual spore produced by many fungi, often for rapid dispersal.
35. **Sporangium:** A specialized structure in which spores are formed and held until release.
36. **Haustorium:** A projection from certain parasitic fungi that penetrates host cells to extract nutrients.
37. **Appressorium:** A flattened, adhesive structure formed by some fungi to attach to and breach the host's surface.
38. **Pathogenicity:** The inherent ability of a fungus to cause disease in a host.
39. **Virulence:** The degree or severity of disease caused by a particular fungal pathogen.
40. **Host Range:** The spectrum of plant species susceptible to infection by a specific fungus.
41. **Disease Cycle:** The sequence of events involved in the development and spread of a fungal disease.
42. **Signs:** The visible evidence of the fungal pathogen itself, such as spores or mycelial growth on the plant.
43. **Symptoms:** The visible changes in a plant's appearance or function due to fungal infection.
44. **Inoculum:** The part of the fungus (spores, mycelium, etc.) that initiates infection.
45. **Infection Court:** The part of the plant where the fungus first enters and establishes itself.
46. **Biotroph:** A fungus that derives nutrients from living host cells.
47. **Necrotroph:** A fungus that kills host cells to obtain nutrients.
48. **Hemibiotroph:** A fungus that initially behaves as a biotroph, then switches to a necrotrophic lifestyle.
49. **Virus:** A microscopic infectious agent that can only replicate inside the living cells of other organisms having a protein coat and nucleic acid.
50. **Virion:** The complete, infective form of a virus outside a host cell, with a core of RNA or DNA and a capsid.

51. **Capsid:** The protein coat of a virus that encloses and protects its genetic material.
52. **Genome:** The complete genetic material (either RNA or DNA. of a virus.
53. **Vector** An organism, often an insect, that transmits a virus from one host to another.
54. **Transmission:** The process of spreading a virus from one host to another.
    - **Mechanical Transmission:** The spread of a virus through direct contact or injury.
    - **Vector Transmission:** The spread of a virus by insects, nematodes, fungi, etc.
    - **Seed Transmission:** The spread of a virus through infected seeds.
55. **Host Range:** The range of plant species that a particular virus can infect.
56. **Symptoms:** The visible changes in a plant's appearance or function due to virus infection.
57. **Mosaic:** The appearance of mottled patterns of light and dark green on leaves.
58. **Stunting:** Reduced growth of the plant.
59. **Chlorosis:** Yellowing of leaves.
60. **Necrosis:** Death of plant tissue.
61. **Systemic Infection:** A virus infection that spreads throughout the entire plant.
62. **Local Infection:** A virus infection that is limited to a specific area of the plant.
63. **Latent Infection:** A virus infection that is present in the plant but does not show symptoms.
64. **Resistance:** The ability of a plant to prevent or limit virus infection.
65. **Tolerance:** The ability of a plant to withstand virus infection without significant yield loss.
66. **Cross-Protection:** A phenomenon where infection by a mild strain of a virus can protect against infection by a more severe strain.
67. **Bacteria:** Microscopic, single-celled organisms that exist in diverse environments.
68. **Prokaryote:** A cell lacking a nucleus and other membrane-bound organelles; bacteria are prokaryotes.
69. **Morphology:** The shape of a cell (e.g., coccus, bacillus, spirillum).
    - **Coccus:** Spherical shape
    - **Bacillus:** Rod shape
    - **Spirillum:** Spiral shape
70. **Gram Stain:** A staining method used to classify bacteria based on cell wall structure.
    - **Gram-positive:** Bacteria that retain the crystal violet stain and appear purple.
    - **Gram-negative:** Bacteria that do not retain the crystal violet stain and appear pink.

71. **Flagellum (plural flagella:** A whip-like appendage used for bacterial movement. **Pilus (plural pili)** Hair-like appendages on the surface of bacteria, involved in attachment and genetic transfer.
72. **Endospore:** A dormant, tough, and non-reproductive structure produced by some bacteria for survival in harsh conditions.
73. **Antibiotic:** A substance that can kill or inhibit the growth of bacteria.
74. **Antibiotic Resistance:** The ability of bacteria to withstand the effects of antibiotics.
75. **Culture:** The process of growing bacteria in a controlled laboratory environment.
76. **Colony:** A visible mass of bacteria growing on a solid surface.
77. **Biofilm:** A community of bacteria that adhere to a surface and produce a protective matrix.
78. **Quorum Sensing:** The ability of bacteria to communicate with each other using chemical signals.
79. **Plasmid:** A small, circular piece of DNA that can replicate independently of the bacterial chromosome.
80. **Conjugation:** The transfer of genetic material between bacterial cells through direct contact.
81. **Binary Fission:** The most common form of asexual reproduction in bacteria, where a single cell divides into two identical daughter cells.
82. **Generation Time:** The time it takes for a bacterial population to double in size.
83. **Growth Curve:** A graph that depicts the different phases of bacterial growth in a population over time.
    - **Lag Phase:** Period of slow growth as bacteria adapt to their environment.
    - **Exponential (Log) Phase:** Period of rapid, exponential growth.
    - **Stationary Phase:** Period of no net growth as resources become limited.
    - **Death Phase:** Period of decline as bacteria die off.
84. **Asexual Reproduction:** Reproduction that does not involve the fusion of gametes.
    - **Binary Fission:** As mentioned above.
    - **Budding:** Formation of a new individual from an outgrowth or bud on the parent cell.
    - **Fragmentation:** Breakage of the parent cell into fragments, each capable of developing into a new individual.
85. **Conjugation:** Transfer of DNA through direct cell-to-cell contact.
86. **Transduction:** Transfer of DNA by bacteriophages (viruses that infect bacteria..
87. **Transformation:** Uptake of free DNA from the environment.
88. **Pattern Recognition Receptors (PRRs):** Plant receptors that recognize PAMPs.

89. **Effector-Triggered Immunity (ETI):** A specific type of plant immune response triggered by the recognition of pathogen effectors.
90. **Effector:** A protein secreted by a pathogen that can manipulate host cell processes.
91. **Avirulence (Avr) Gene:** A gene in a pathogen that encodes an effector that triggers a defense response in plants with a corresponding resistance (R) gene.
92. **Resistance (R) Gene:** A gene in a plant that encodes a receptor that recognizes a specific pathogen effector and triggers a defense response.
93. **Gene-for-Gene Hypothesis:** A model that explains how plants with specific R genes recognize pathogens with corresponding Avr genes, leading to disease resistance.
94. **Quantitative Resistance (QR):** Resistance that is controlled by multiple genes and is often partial or incomplete.
95. **Qualitative Resistance:** Resistance that is controlled by a single gene and is often complete or absolute.
96. **Systemic Acquired Resistance (SAR):** A long-lasting, broad-spectrum resistance response that is triggered by a localized infection.
97. **Induced Systemic Resistance (ISR):** A form of induced resistance that is mediated by beneficial microbes.
98. **Cultural Control:** The use of farming practices to reduce disease pressure, such as crop rotation, planting resistant varieties, sanitation (removal of infected plant debris), and proper irrigation and fertilization.
99. **Biological Control:** The use of living organisms, such as beneficial fungi, bacteria, or insects, to suppress plant diseases by antagonizing or outcompeting the pathogens.
100. **Chemical Control:** The use of pesticides, such as fungicides, bactericides, and nematicides, to kill or inhibit the growth of pathogens. However, this method should be used judiciously to avoid negative impacts on the environment and the development of resistant pathogens.

12

Practice Questions

1. Tea rust is caused by
   - a. MLO
   - b. Virus
   - c. Bacteria
   - d. Algae
2. The *Colletotrichum falcatum* produces
   - a. Zygospores
   - b. Oospores
   - c. Ascospores
   - d. Conidiospores
3. For the effective control of wilt, pigeon pea should be intercropped with
   - a. Sorghum
   - b. Pearl millet
   - c. Maize
   - d. Mung
4. The most important symptom of wilt of cotton (caused by *Fusarium oxysporum* f.sp. *vasinfectum*) is
   - a. Necrosis
   - b. Yellowing of tissues
   - c. Discolouration of tissues and plugging of vessels by hyphae
   - d. All of these
5. Wilt disease of sugarcane was first time reported in India from
   - a. Punjab
   - b. Tamil Nadu
   - c. Bihar
   - d. Uttar Pradesh
6. The pathogen responsible for charcoal rot of soybean is
   - a. *Ascochyta rabiei*
   - b. *Macrophomina phaseolina*
   - c. *Rhizoctonia solani*
   - d. *Penicillium chrysogenum*
7. The incidence of black scurf of potato is more in
   - a. Sandy soil
   - b. Clay soil
   - c. Alluvial soil
   - d. Loam soil
8. The attack of sheath blight of rice (caused by *Rhizoctonia solani*) is more during
   - a. Germination
   - b. Active tillering stage
   - c. Flowering
   - d. All of these
9. Resistance to broad range of Powdery mildew is offered by
   - a. RPW8
   - b. cf-8
   - c. Hm-1
   - d. Xa21
10. Differential resistance is also called as
   - a. Horizontal resistance
   - b. Vertical resistance
   - c. Apparent resistance
   - d. Field resistance

11. Supressors of RNAi silencing
    a. HC-component of Poty viruses
    b. p25 protien of potato virus X
    c. Both a and b
    d. None
12. RISC is formed by
    a. ssRNA+ Proteins
    b. Proteins
    c. ssRNA
    d. dsRNA
13. Viroids were discovered by
    a. Diener
    b. Prusiner
    c. Doi
    d. Kassanis
14. Pathogenicity of bacteria is controlled by
    a. Type-III
    b. Type-II
    c. Type-I
    d. Type-IV
15. Mycoplasma cell membranes have how many layers
    a. One layer
    b. Two layers
    c. Three layers
    d. Four layers
16. Leaf puckering as a result of different growth rates in adjoining tissue is
    a. Savoying
    b. Hyperplasia
    c. Sarcody
    d. Intumesence
17. The uredospores of *Puccinia graminis* are disseminated by
    a. Wind
    b. Animals
    c. Insect
    d. Birds
18. G + C content of prokaryotes is
    a. 20-77%
    b. 30-50%
    c. 50-60%
    d. 70-80%
19. Bacterial cell division mainly by
    a. Binary division
    b. Fragment
    c. Budding
    d. None of these
20. Shape of MLO is
    a. Pleomorphic
    b. Cuboid
    c. Circular
    d. Polymorphic
21. Sensitivity of ELISA
    a. 0.01-0.1 ng/ml
    b. 10-20 ng/ml
    c. 0.2-1 ng/ml
    d. 5-20 ng/ml
22. Teichoic acid found in
    a. Fungi
    b. Gram + ve bacteria
    c. Gram - ve bacteria
    d. Protozoa
23. Father of bacteriology
    a. Louis Pasteur
    b. Robert Koch
    c. Doi
    d. E.F.Smith

24. Plant Pathology book was written by
   a. R.S.Singh b. Agrios
   c. V.S.Singh d. A.P.Sinha
25. Rickettsia disease is caused by
   a. Gram –ve obligate bacteria b. Gram +ve fastidious bacteria
   c. *Clavibacter xyli* d. Both b and c
26. In north India and central India, the black rust inoculum come from
   a. South b. Hilly area
   c. From USA d. From Netherlands
27. Teliospore of rust has germ pores in number
   a. 1-4 b. 2-5
   c. 3-6 d. 4-8
28. Rice blast pathogen perfect stage is
   a. *Pyricularia oryzae* b. *Magnaporthe grisea*
   c. *Helminthosporium oryzae* d. *Rhizoctonia solani*
29. Smut of maize is caused by
   a. *Ustilago tritici* b. *Ustilago maydis*
   c. *Ustilago hordei* d. None of these
30. Virus capsid is made up of
   a. Protein b. Carbohydrate
   c. Lipid d. Nucleic acid
31. Phyllody disease of sesamum spread by
   a. Leaf hopper b. Jassid
   c. Aphid d. Whitefly
32. Mycoviruses contain
   a. ds RNA b. ds DNA
   c. ss RNA d. ss DNA
33. Hrp genes were discovered by
   a. Albersheim b. Lindgren & Peet
   c. Morris d. Staskawicz
34. Microorganisms in Diseased Plants is written by
   a. R.S.Singh b. C.D.Mayee
   c. S. Nagarajan d. A.Mahadevan
35. Who recognized hypersensitive response in bacteria
   a. Stakmann b. Z Klement
   c. H.M.Ward d. Biffen
36. Downy mildew of pea caused by
   a. *Perenospora pisi* b. *Albugo candida*
   c. *Erysiphe polygoni* d. None of these

37. Father of Indian Mycology
    a. E.J.Butler b. K.C.Mehta
    c. Mundkur d. R.S.Singh
38. Renowned scientist Norman E. Borlaug belongs to
    a. Agronomy b. Genetics
    c. Entomology d. Plant Pathology
39. Which pathological scientist got Borlaug award
    a. Rattan lal b. V.L.Chopra
    c. C.D.Mayee d. S.Nagarajan
40. Which of the following bacteria generally used in genetic engineering for transfer of character from one organism to another
    a. *Bacillus thuringinesis* b. *Pseudomonas syringe*
    c. *Rhizobium meliloti* d. *Agrobacterium tumefaciens*
41. Micrografting is used to produce plants free from
    a. Virus b. RLO's
    c. MLO's d. Bacteria
42. Black heart is a physiological disorder of
    a. Tomato b. Chilli
    c. Potato d. Cabbage
43. Which of the following is not correctly matched?
    a. Mustard- white rust b. Paddy-Brown rust
    c. Wheat-Red rust d. Groundnut – Tikka
44. Tungro disease of rice is spread by
    a. *Nephotettix virescense* b. *Sogatella fucifera*
    c. *Nilaparvatha leugens* d. *Thrips tabaci*
45. The term hypersensitive response is given by
    a. Stakmann b. Erickson
    c. H.M.Ward d. W.A.Orton
46. Soil microorganism are most active at
    a. 18-20 ^{0}C b. 26-28 ^{0}C
    c. 30-30 ^{0}C d. 34-36 ^{0}C
47. VAM is
    a. Bacteria b. Fungi
    c. Virus d. Algae
48. Which of the following is not a disease of apple?
    a. Bird eye b. Powdery mildew
    c. Downy mildew d. Bunchy top
49. Buck eye rot is disease of which crop
    a. Water chestnut b. Sweet potato
    c. Pods of garden pea d. Tomato fruits

50. The major storage fungi that affects the food grain is
    a. *Rhizobium* b. *Mucor*
    c. *Cercospora* d. *Aspergillus*
51. Yellow leaf mosaic of Okra spreads by
    a. Jassids b. Borers
    c. Jassid and borers d. Whiteflies
52. Polymorphism in fungi was first noticed by
    a. Anton de Bary b. Tulsane brothers
    c. E.M.Fries d. Muller
53. Which of the disease (s) induce floral abnormalities?
    a. Downy mildew of mustard b. Green ear of bajra
    c. White rust of crucifers d. Both b. and c.
54. Which one of the following microbial agents is being commercially exploited as biocontrol agent?
    a. *Bacillus subtilis* b. *Penicillium notatum*
    c. *Sclerotium rolfsi* d. *Trichoderma viride*
55. Panama disease of banana is prevented by
    a. Spraying zinc carrier b. Spraying copper fungicide
    c. Application of lime to the soil d. Providing adequate irrigation
56. Which one the following fungicides is not systemic in nature
    a. Vitavax b. Thiram
    c. Benlate d. Topsin
57. Which one of the following is/are entomopathogenic fungi?
    a. *Beauveria bassiana* b. *Metarrhizium anisopliae*
    c. *Trichoderma harzianum* d. Both a. and b.
58. Which penicillium species used in Cheese making
    a. *Penicillium notatum* b. *Penicillium roqueforti*
    c. *Penicillium diversicaudatum* d. *Penicillium crysogenum*
59. CARNA-5
    a. Satellite RNA b. Satellite Virus
    c. Virus d. Bacteria
60. Bacterial diseases are controlled by use of chemicals
    a. Kelthane b. Fungicide
    c. Antibiotics d. Viricides
61. Application of potash increases
    a. Resistance to water logging b. Frost resistance in plants
    c. Disease resistance in plants d. None of these
62. Decomposition of organic matter in submerged soil is carried out by
    a. Bacteria b. Actinomycetes
    c. Fungi d. Earthworm

63. The sedimentation coefficient for most of viruses is between
    a. 50-200S b. 100-200S
    c. 200-300S d. 25-50S
64. Most of viruses have buoyant density between
    a. 1.2 and 1.6 g/cm3 b. 1.4 and 1.8 g/cm3
    c. 1.0 and 1.2 g/cm3 d. 1.8 and 2 g/cm3
65. Phenols present in the red scales of onion
    a. Catechol b. Procatechoic acid
    c. Saponin d. a&b
66. Alteration of generation is seen in
    a. *Alternaria* b. *Allomyces*
    c. *Neovassia* d. *Tilletia*
67. Local and systemic infection is seen in
    a. White rust of crucifers b. Late blight of potato
    c. Coffee rust d. Rice blast
68. Viruses contain
    a. RNA b. DNA
    c. Both RNA and DNA d. Either RNA or DNA, never both
69. The viruses which are usually helped or accompanied by smaller spherical particles of another serologically unrelated virus is known as
    a. Satellite virus b. Gemini viruses
    c. Viroid d. Capsid
70. Select the correct pair from the following
    a. Hypotrophy- decrease in size b. Hypertrophy-increase in size
    c. Hyperplasia- increase in number d. All of these
71. In plants buckling, puckering and blistering symptoms are produced by
    a. Bacteria b. Fungi
    c. Viruses d. Mycoplasma
72. The fungi which transmit plant viruses belong to class
    a. Basidiomycetes b. Oomycetes
    c. Zygomycetes d. Plasmodiophoromycetes
73. For quick and accurate detection of viruses can be done by
    a. ELISA b. HADAS
    c. IEM d. All of these
74. Potato viruses are spread by
    a. Aphids b. Jassids
    c. Nematodes d. Tubermoth

75. Most of the plant viruses transmitted by
   a. Whiteflies   b. Plant hopper
   c. Leaf hopper   d. Aphids
76. Bunchy top of papaya is spread by
   a. Nematodes   b. Fungi
   c. Whitefly   d. Leaf hoppers
77. Yellow mosaic of legumes was first reported in India from
   a. Shimla   b. Solan
   c. Delhi   d. Kanpur
78. Potato spindle tuber disease is transmitted by
   a. Mechanically   b. biologically
   c. Water   d. All of these
79. Select the correct pair from the following matched items
   a. Holo stem parasite- Cuscuta reflexa   b. Stem semi parasite-Mistle toe
   c. Holo root parasite – Orobanche   d. All of these
80. Suicidal germination takes place in
   a. Dodder   b. Striga
   c. Loranthus   d. Dendrophthoe falcata
81. Chitin is composed of
   a. Chitinin   b. Glucose
   c. Cerasan   d. N-acetyl glucosamine
82. Damping off and leaf blights are very effectively checked by
   a. Bordeaux mixture   b. Burgundy mixture
   c. Thiram   d. Copper oxychloride
83. The organo-mercurials fungicides used for dry seed treatment usually contain mercury
   a. 0.5%   b. 1.0 %
   c. 2.0 %   d. 4.0%
84. Dinocap (methyl heptyl dinitrophenyl crotonate) is sold in market as
   a. Bravo   b. Dexon
   c. Botron   d. Karathane
85. Von Schleming and Kulka first time demonstrated the systemic activity of fungicides in
   a. 1956   b. 1966
   c. 1972   d. 1976
86. Select the fungicides
   a. Pyrimidines   b. Triazoles
   c. Metalaxyl   d. All of these

87.Infective spores in rust life cycle are

a. Uredospores and Aeciospores
b. Uredospores and Teliospores
c. Teliospores and Basidispores
d. Pycniospores and Uredospores

88. Tick out he pair which is wrongly matched

a. Oxycarboxin- Plantvax
b. Carboxin- Benlate
c. MBC-Bavistin
d. Benomyl- Benlate

89. Most commonly used transposon in molecular plant pathology is

a. Bn6
b. Hm1
c. Tn5
d. T4

90. Black or Silvery scurf of potato is caused by

a. *Sclerotium rolfsii*
b. *Rhizoctonia solani*
c. *Ralstonia solanacearum*
d. *Phytophthora infestans*

91. Forecasting system developed for apple scab on the basis of

a. Amount of inoculum
b. Leaf wetness period and temperature
c. RH
d. Temperature

92. Largest know viroid

a. Citrus Exocortis viroid
b. Potato spindle tuber viroid
c. Coconut cadang cadang viroid
d. Apple scar viroid

93. The pathogenicity of a viroid is determined by which region

a. Pathogenicity region & Terminal region
b. Terminal region & Variable region
c. Central conserved region & Pathogenicity region
d. Variable region

94. The science which deals with cause of the plant diseases and nature of the causal agent

a. Fertilization
b. Etiology
c. Aetiology
d. Both b. and c.

95. The sexual, thick walled and resting spores of the rust fungi is known as

a. Basidiospores
b. Urediospores
c. Teliospores
d. Aeciospores

96. Who discovered the downy mildew for the first time in India?

a. E.J.Butler
b. J.F.Dastur
c. B.B.Mundkur
d. K.R.Kirtikar

97. Which one of the following is a single cell fungi?

a. Yeast
b. Aspergillus
c. Penicillium
d. Alternaria

98. What is the source of Agar Agar?

a. Bacteria b. Fungi

c. Mycoplasma d. Algae

99. Who is the father of Forest pathology

a. Robert Hartig b. Butler

c. Agrios d. Kuhn

100. Who is the father of modern plant pathology

a. Robert Hartig b. Butler

c. Stalkman d. De Bary

Answers

1 d	2 d	3 a	4 c	5 c	6 c	7 b	8 b	9 a	10 a
11c	12 c	13 a	14 a	15c	16a	17 a	18 a	19 a	20 a
21a	22 b	23. b	24. b	25. a	26. b	27. a	28. b	29. b	30. a
31a	32 a	33a	34 a	35 b	36 a	37 a	38 a	39 d	40 d
41a	42 c	43 c	44 a	45 a	46 b	47 b	48 d	49 d	50 d
51d	52 d	53d	54 d	55 c	56b	57 d	58 b	59 c	60c
61c	62 a	63 c	64 b	65 d	66 b	67 a	68d	69 a	70 d
71c	72 d	73 a	74 a	75 d	76 d	77 c	78 a	79 d	80 b
81 d	82 d	83 b	84 d	85 b	86 d	87 a	88b	89 c	90 b
91 b	92 a	93 a	94 b	95 c	96a	97 a	98 d	99 a	100 d

Multiple Choice Questions

1. Haustoria are produced.

a. By fungi parasitising animal tissues b. By hypha of obligate parasite

c. In artificial cultures d. After the hyphae become intercellular

2. The common food stored in fungi is

a. Glycogen b. Starch

c. Oil d. Fat

3. Stroma is

a. Compact somatic hyphae with fruit bodies b. loosely interwoven hyphae

c. A small hyphae branch d. A group of spores

4. All fungi are

a. Autotroph b. Heterotrophs

c. Saprophytes d. Parasites

5. A sclerotium produced by fungi is

a. An underground structure b. A hard resting body

c. Mainly food storing organ d. Easily carried off by wind

6. Somatogamy is The
   a. Fusion of gametes
   b. Fusion of vegetative cells
   c. Contact between two gametes
   d. Copulation between gametangia

7. Myxomycetous fungi are also called
   a. Slime molds
   b. Bread molds
   c. Black molds
   d. Blue molds

8 The somatic phase in Myxomycetousfungus is called
   a. Hyphae
   b. Mycelium
   c. Plasmodium
   d. None of the above

9. The zoospore in synchytrium is
   a. Posteriorly uniflagelated
   b. Anteriorly uniflagelated
   c. Anteriorly biflagelated
   d. Laterly biflgelated

10. The sexual reproduction in lower fungi is usually
   a. Planogametic fusion
   b. Gametangial contact
   c. Somatogamy
   d. None

11. Wart disease of potato is caused by
   a. *Syncytrium endobioticum*
   b. *Alternaria solani*
   c *Pytophthora infestans*
   d. *Puccinia graminis*

12. Which of the following has two typesof sporangia in their life cycle
   a. *Pythium*
   b. *Allomyces*
   c. *Monoblepharis*
   d. *Albugo*

13. Motile asexual reproductive bodies arecalled
   a. Chlamydospores
   b. Zoospores
   c. Aplanospores
   d. Zygospores

14. The laterally bi-flagellated zoospores are usually
   a. Pear shaped
   b. Kidney shaped
   c. Lemon shaped
   d. Irregular in shape

15. *Plasmodiophora brassica* causes a diseases in crucifers which is known as
   a. Club root
   b. White rust
   c. Early blight
   d. Late blight

16. The zygote in the life cycle ofPlasmodium is
   a. Uniflagelate
   b. Biflagellate
   c. Non-flagellate
   d. Quadriflageallate

17. Haustoria are produced by
   a. Fungal parasites of animal
   b. Intercellular hyphae of Obligate parasites
   c. Fungi growing on artificial media
   d. Intercellular hyphae of parasiticfungi

18. Which of the following disease iscaused by Plasmopara?
    a. Green ear disease of bajra
    b. Late blight of potato
    c. Downy mildew of grapes
    d. Damping off of seedlings
19. Damping off of seedling is caused by
    a. *Albugo candida*
    b. *Pythium debaryanum*
    c. *Sclerospora graminicola*
    d. *Peeronospora parasitica*
20. *Albugo candida* is an example of
    a. Obligate parasite
    b. Obligate saprophyte
    c. Facultative sapropyte
    d. Facultative parasite
21. Which of the following causes whitespots on leaves and hypertrophied floral parts.
    a. *Phytophthora infestans*
    b. *Albugo candida*
    c. *Pythium debaryanum*
    d. *Sclerospora graminicola*
22. In Phytophothra asexual reproductive bidies behave
    a. Conidia
    b. Conidiosporangia
    c. Sporagia
    d. Both conidia and conidiosporangia
23. White rust of crucifers is Pseudorust because
    a. Because causal organism is an oomycete
    b. The colour of pustule is not brown
    c. The disease occurs in members of family cruciferae
    d. The disease has not been observed on wheat
24. The production of two differenttypes of zoospores is termed as
    a. Cogugation
    b. Autoaecium
    c. Heterothalism
    d. Diplanatism
25. Sporangiophores of Plasmopara arebranched at
    a. Acute angles
    b. Obtuse angle
    c. Sparsly branched
    d. Right angles
26. In *Albugo candida* conidiosporagiaare produced in
    a. Acropetal succession
    b. Basipetal succession
    c. Longitudinal manner
    d. Irregular manner
27. Thick walled resting spores producedby members of Peronosporales are called
    a. Oospores
    b. Ascopsores
    c. Zygospores
    d. Zoospores
28. Father of Plant Pathology is
    a. Butler
    b. Anton De Bary
    c. K C Mehta
    d. BB Mundkar
29. Heterothalism was discovered by
    a. E. J. Buttler
    b. J.H. Craigie
    c. A.F.Blakeslee
    d. H.C. Dube

30. The mycelium is coenocytic in thegenus
    a. Mucor
    b. Synchytrium
    c. Plasmodiophora
    d. Physarum
31. Which of the following is the result ofsexual reproduction
    a. Zygospore
    b. Zoospore
    c. Conidia
    d. Chlamydospore
32. Thick walled resting spores formedasexually are known as
    a. Oospore
    b. Zygospore
    c. Oidospore
    d. Chlamydospore
33. The fungi, which have many commonfeatures with algae especially in the thallus structure and reproduction, have been placed in class
    a. Phycomycetes
    b. Ascomycetes
    c. Basidiomycetes
    d. Deuteromycetes
34. The Sporangiophores of *Rhizopus stolonifer* are produced
    a. Directly above the rhizoids
    b. From any point on the aerial stolon
    c. From the center of the stolon
    d. From the root cells
35. Collumellate sporangia arecharacteristic feature of
    a. Synchytrium
    b. Rhizopus
    c. Achyla
    d. Albugo
36. A fungus producing 8 spores ina saclike structure should rightly be placed in
    a. Basidiomycetes
    b. Ascomycetes
    c. Phycomycetes
    d. Deuteromycetes
37. A closed ascocarp is named as
    a. Perithecium
    b. Apothecium
    c. Cleistothecium
    d. Pseudothecium
38. Ascocarps are the characteristicfeature of
    a. Ascomycetes
    b. Deuteromycetes
    c. Basidiomycetes
    d. Deuteromycetes
39. Stroma is
    a. Compact somatic hyphae with fruit
    b. Loosely interwoven hyphae bodies
    c. Small hyphal branch
    d. A group of spores
40. The protective covering of sterile hyphae around an ascocarp is known as
    a. Periderm
    b. Peridium
    c. Appendages
    d. Epiderm
41. The asexual reproduction in ascomycetes is generally through
    a. Conidia
    b. Zygospore
    c. Zoospore
    d. Aplanospore

42. An ascocarp which is flask shapedwith an ostiole in the center is
    a. Apothecium b. Perithecium
    c. Cleistothecium d. Hypothecium
43. Perfect stage of fungus implies that
    a. The fungus is perfectly healthy b. The fungus is reproduce asexually
    c. The fungus reproduces sexually d. The fungus reproduces restingspores
44. Ascospores are produced in
    a. basidium b. ascus
    c. sporangium d. gametangium
45. Asexual reproductive structures formed directly at the tip of hyphaeare
    a. spores b. Aplanospores
    c. Conidia d. Zygospore
46. In Penicillium, the conidia areproduced
    a. In sori b. On unbranched conidiophores
    c. In branched conidiophores d. On both branched and unbranched conidiophores
47. Penicillium was discovered by
    a. Ian Fleming b. Alexander Fleming
    c. E. Jenner d. Louis Pasteur
48. Which of the following disease iscaused by ***Aspergillus***
    a. Cholera b. *Aspergillosis*
    c. Tetanus d. Otomycosis
49. The basal cell from where the conidiophore of ***Aspergillus*** arises isknown as
    a. Trophocyst b. Tip cell
    c. Foot cell d. Terminal cell
50. The reproductive organ of *Aspergillus* is
    a. Antheridium and oogonium b. Spermatium and oogonium
    c. Antheridium and ascogonium d. None
51. One of the important characters of *Hemiascomycetidae* **is**
    a. Asci produced in cleistothecium b. Asci produced in perithecium
    c. Asci produced in apothecium d. Asci are naked
52. Naked asci ae produced in Taphrinia Mucor
    a. Taphrinia b. Mucor
    c. *Aspergillus* d. *Penicillum*
53. Latin binomial of brewers yeast is
    a. *Rhizopus stolonifer* b. *Aspergillus flavus*
    c. Saccharomyces cerevisae d. *Penicillum notatum*

54. Fermentation of sugars occurs by
    a. Mucor b. Saccharomyces
    c. Aspergillus d. Penicillum
55. *Taphrina deformans* causes
    a. Rusts of wheat b. Peach leaf curl
    c. Stem gall of coriander d. White rust of crucifers
56. Powdery mildew disease is caused by
    a. *Erysiphe polygoni* b. *Sclerospora graminicola*
    c. *Neurospora sp.* d. *Claviceps purpurea*
57. The fungus, which is important for its use in genetic studies
    a. Aspergillus b. Rhizopus
    c. Penicillum e. Neurospora
58. Ergot is extracted from
    a. *Erysiphae polygoni* b. *Phyllactinai sp.*
    c *Calviceps purpurea* d. Synchytrium endobiticum
59. The characteristic of cleistothecium of Phyllactina is that the appendagesare
    a. Dicholomously branched b. Curled at apex
    c. Provided with bulbous base d. Absent
60. In Claviceps the ascocarp is
    a. Perithecium b. Cleistothecium
    c. Apothecium d. Hypothecium
61. Cup shaped ascocarp are characteristic features of
    a. Plectomycetes b. Pyrenomycetes
    c. Discomycetes d. Ascomycetes
62. Which of the following is corpophilous fungus
    a. Phytophthora b. Pilobolous
    c. Ascobolus d. Aspergillus
63. Which of the following is edible fungus
    a. *Morchella deliciosa* b. *Pyronema confluens*
    c. *Aspergillus nidulans* d. *Phytophthora infestans*
64. Asci are produced alternating with
    a. Elaters b. Paraphysis
    c. Mycelium d. Nothing
65. Fertile layer in pileus of Morchella iscalled
    a. Hymenium b. Subhymenium
    c. Trama d. Cortex
66. Apothecium in Peziza is
    a. Aerial and short lived b. Underground and short lived
    c. Aerial and persistant d. Underground and persistant

67. Which of the following spores are produced by Morchella?
    a. Oospore
    b. Ascospore
    c. Basidiospore
    d. Chlamydospores
68. The number of basidiospore produced on each basidium is
    a. 2
    b. 4
    c. 6
    d. 8
69. Basidiospores are characteristic spores of class
    a. Phycomycetes
    b. Ascomycetes
    c. Basidiomycetes
    d. Deuteromycetes
70. The process of dikarotization involves
    a. plasmogamy only
    b. plasmogamy associated withtransfer of nuclei
    c. karyogamy
    d. plasmogamy and karyogamy
    e. spermatium and ascogonium
71. Basidiospores are different from Ascospores because the former areproduced
    a. Externally
    b. Internally
    c. Both externally and internally
    d. Neither externally and internally
72. Dolipore septum is characteristicfeature of
    a. Ascomycetes
    b. Basidiomycetes
    c. Discomycetes
    d. Phycomycetes
73. Secondary dikaryotic mycelium is identified on the basis of presence of
    a. Clamp connection
    b. Antherida
    c. Oogonia
    d. Ascogonia
74. Whip smut disease of sugar cane iscaused by
    a. *Ustilago scitaminea*
    b. *Ustilago maydis*
    c. *Ustilago hordei*
    d. *Ustilago nuda*
75. Smuts are regarded as
    a. Obligate parasites which can be cultured on axenic culture media
    b. Obligate parasites which can becultured on synthetic culture media
    c. Obligate parasite which can notbe cultured at all
    d. Obligate parasite which caninfect only living host
76. Loose smut of wheat is caused by
    a. *Albugo candida*
    b. *Ustilago tritici*
    c. *Puccinia grminis*
    d. *Syncytrium endobioticum*
77. Smut spores act as
    a. Uredospores
    b. Teleutospores
    c. Basidiospores
    d. Aeciospores

78. Production of spore ball is thecharacteristic of
   a. *Tilletia* b. *Ustilago*
   c. *Tolyposporium* d. *Puccinia*
79. Ustilago produces in large quantitiesthe spores known as
   a. Uredospores *b. Zoospores*
   c. Zygospores *d. Teleutospores*
80. Covered smut of barley is caused by
   a. *Ustilago hordei* b. *Ustilago nuda*
   c. *Ustilago scitaminae* d. *Ustilago avenae*
81. Ustilago produces in large quantitiesthe spores known as
   a. Uredospores b. Zoospores
   c. Zygospores d. Teleutospores
82. Rust of gram is caused by
   a. *Pucinia* b. *Uromyces*
   c. *Phragmidium* d. *Urocystis*
83. Common name of the disease causedby the members of Uredinales is
   a. Rusts b. Smuts
   c. White rusts d. Wart
84. In macrocyclic rust, zero stagerefers to
   a. Aecium b. Uredium
   c. Pycnium d. Basidium
85. Repeatedly produce spores in rustcycle called
   a. Aeciospores b. Uredospore
   c. Basidiospore d. Pycnidiospore
86. When a fungus requires two host tocomplete its life cycle, the fungus is referred as
   a. Homothalic *b.* Heterothalic
   c. Autoecious *d.* Heteroecious
87. The West Bengal famine of 1942 wascaused by the infection of
   a. *Puccinia graminis tritici* *b. Helminthosporium oryzae*
   c. *Phytophthora infestans* *d. Synchytirum endobioticum*
88. A macrocyclic rust is one, whichneeds, which
   a. Produces five types of spores to complete its life cycle
   b. Two different hosts to complete its life cycle
   c. Does not show any sexual reproduction
   d. Has a prolonged life cycle
89. When two different hosts are required by a fungus to complete itslife cycle, it is known as
   a. Autoecism b. Heteroecism
   c. Homothalism d. Heterothalism

90. The mushroom is
    a. A plant made up of fine greenthreads
    b. An edible fungus
    c. A bryophyte without roots
    d. A flowering plant
91. Basidia in Agaricus produces on
    a. Gills
    b. Pileus
    c. Stipe
    d. Rhizomorph
92. Toad stool can not manufacture theirown food because
    a. They have no chlorophyll
    b. They have no roots
    c. They have have no leaves
    d. They have no seed for food
93. In which of the following gills arepresent
    a. *Aspergillus*
    b. *Alternaria*
    c. *Agaricus*
    e. *Mucor*
94. Which of the following is an ediblefungus
    a. *Aspegillus*
    b. *Alternaria*
    c. *Agaricus*
    d. *Amanita*
95. Which of the following are woodrotter
    a. *Agaricus*
    b. *Polyporus*
    c. *Morchella*
    d. *Mucor*
96. Micro and macro conidia are presentin
    a. Alternaria
    b. Aspergillus
    c. Mucor
    d. Fusarium
97. Fungi imperfecti are so named because of the absence of
    a. Asexual reproduction
    b. Vegetative reproduction
    c. Sexual reproduction
    d. Reproduction
98. Sexual reproduction is absent in themembers of the class
    a. *Phycomycetes*
    b. *Ascomycetes*
    c. *Plectomycetes*
    d. *Deuteromycetes*
99. Conidiophore and conidia aggregated in pycnidia are present in
    a. *Alternaria*
    b. *Colletotrichum*
    c. *Phyllosticta*
    d. *Fusarium*
100. Nitrifying bacteria is
    a. *Rhizobium*
    b. *Nitrosomonas*
    c. *Nitrobacter*
    d. *Bacillus*
101. Insoluble Copper fungicides are
    a. More phytotoxic than soluble Cu compounds
    b. Used for the control of similar plantdiseases as Bordeaux mixture
    c. Those, which contain more Cu and are not readily soluble in water
    d. Recommended at the rate of 2 lbs/100gallons of water

102. Inorganic Sulphur compounded aremostly used to control
   a. Blight or anthracnose
   b. Apple scab
   c. Brown rot of stone fruits
   d. Powdery mildews

103. Ferbam consists of
   a. Two molecules of dithiocarbamic acid
   b. Three molecules of dithiocarbamicacid
   c. One molecule of dithiocarbamic acid
   d. Four molecules of dithiocarbamic acid

104. Mancozeb is a formulation of
   a. Mancozeb and zinc
   b. Maneb and zinc ion
   c. Manganese and zinc
   d. Maneb and manzate

105. Addition of zinc to maneb
   a. Makes it a broad spectrum fungicide
   b. Reduces phytotoxicity
   c. Improves fungicidal properties
   d. All of these

106. Pentachlorobenzene is mainly effectiveagainst
   a. *Pythium*
   b. Vegetable diseases
   c. *Rhizoctonia*
   d. Seed-borne diseases of ornamentals

107. Quinones are
   a. Naturally occurring substances in fungi
   b. Plant phenolic compounds
   c. Toxic to microbes
   d. Associated with resistance to diseases

108. Benlate is mainly effective against
   a. Oomycetes
   b. Downy mildews
   c. Powdery mildews
   d. Pythium and phytophthora

109. Stripe rust of wheat is favored by
   a. Cool moist weather
   b. Warm humid weather
   c. Warm dry weather
   d. Hot humid weather

110. Early blight of potato occures
   a. Early in the season whentemperature is low
   b. Early in the season whentemperature is moderate
   c. Early in the season whentemperature is high
   d. Early in the season whentemperature is cool

111. *Puccinia graminis tritici* survives adverse environmentalconditions in the form of
   a. Dormant mycelium
   b. Uredospores
   c. Teliospores
   d. Basidiospore and pycniospore

112. *Fusarium oxysporum* f. sp.*ciceri* can be controlled by
   a. Spraying fungicides
   b. Cultivation of a tolerant variety
   c. Using disease free seed forsowing
   d. All of the above
113. *Colletotrichum falcatum* is
   a. Facultative
   b. Saprophyte
   c. Facultative parasite
   d. All of these
114. Micron is
   a. One million of a centimeter
   b. One thousand of a meter
   c. One thousand of millimeter
   d. One hundred of a micrometer
115. Ear cockle of wheat can becontrolled by
   a. Seed treatment with captan
   b. Proper cultural practices
   c. Using sterilized soil
   d. All of these
116. *Ascochyta rabiei* survives in the soilin the form of
   a. dormant mycelium
   b. Resting sporangia
   c. Conidia
   d. *Pseudothecia*
117 A virusoid is
   a. A virus particle consisting of RNA
   b. Single stranded small circularnaked RNA
   c. Virus particle without protein coatreplicating in host nucleus
   d. Subviral circular RNA componentof some RNA viruses
118. The suitable temperature for the development of citus canker is
   a. 10-15 °C
   b. 12-16 °C
   c. 20-30 °C
   d. 30-35 °C
119. New bunt of wheat is
   a. Seedling penetration type
   b. Blossom penetration type
   c. Partial bunt
   d. Systemic penetration type
120. Mistletoe is a parasite of
   a. Roots
   b. Flowers
   c. Stem
   d. Leaves
121. *Ustilago tritici* perpetuates in
   a. The soil
   b. The seed
   c. On the seed
   d. All of these
122. *Anguina tritici* gets contaminated with *Corynebacterium tritici* in the
   a. Spikelets
   b. Soil
   c. Seed
   d. All of these
123. *Taphrina deformans* survivesitsadverse environmental conditions in the form of
   a. Ascospore
   b. All of the above
   c. Resting sporangia
   d. Dormant mycelium

124. CLCuV can be managed by
   a. Cultivation of resistantvariety
   b. Control of whitefy
   c. Seed treatment
   d. All of the above
125. Most of the triazole fungicides are used as
   a. Protectants
   b. Eradicants
   c. Ergosterol biosynthesis inhibitors
   d. Against many ascomycetes andbasidiomycetes
126. Systemic fungicides are
   a. Multisite inhibitors
   b. Translocated in one direction only
   c. Used mostly as eradicants
   d. Specific site inhibitors
127. Ethazole is effective against
   a. Sclerotium
   b. Air-borne diseases
   c. Fusarium when combined with PCNB
   d. Fusarium and Rhizoctonia
128. Resistant strains of plant pathogenshave appeared against
   a. Bayleton
   b. Ridomil
   c. Nimrod
   d. Tilt
129. Oxathiins are
   a. Formulation of Carboxin andBlitox-50
   b. Effective against damping off andsmuts
   c. Used mainly as seed treatment
   d. Inhibitors of mitochondrial respiration
130. Tetracyclines are not effective against
   a. Plant pathogenic bacteria
   b. Plant pathogenic bacteria and mycoplasma
   c. Fastidious bacteria
   d. Fungi
131. Agrimycin-100 is a mixture of
   a. Streptomycin and terramycin
   b. Streptomycin sulphate and terramycin
   c. Streptomycin and tetracycline
   d. Chlortetracycline and tetracycline
132. Carbofuran is a member of
   a. Carbamates
   b. Heterocyclic carbamates
   c. Dialkyldithiocarbamate
   d. Monoalkydithiocarbamate
133. Mites accompanied with powderymildews can also be controlled by
   a. Captafol
   b. Mildex and Phygon
   c. Dinocap
   d. Karathane and Benlate
134. The Cu for the control of plantdiseases was first used by
   a. Prevost
   b. Grison
   c. Millardet
   d. Schulthess

135. Malathion is
   a. Systemic
   b. Systemic contact
   c. Pseudosystemic
   d. Short-term contact
136. Which one of the following is not usedfor the control of rust of beans and wheat
   a. Dithane M-20
   b. Dithane R-14
   c. Dithane D-14
   d. Dithane A-40
   e) Dithane Z-78
137. Which one of the following is not afungicide
   a. Tilt
   b. Azoxystrobin
   c. UV light
   d. Chloropicrin
138. A typical plant virus will contain
   a. RNA
   b. DNA
   c. RNA OR DNA
   d. None
139. Viroids possess
   a. One capsid protein
   b. Many additional coat
   c. RNA
   d. RNA and Protein coat
140. Phytoplasmas are sensitive to
   a. Penicillin
   b. Streptomycin
   c. Not sensitive to antibiotics
   d. Copper Sulphate
141. Tobacco Mosaic Virus wasdiscovered by
   a. Stanley
   b. Mayer
   c. Ivanowski
   d. Beijerinck
142. Virus diseases can be controlled by
   a. Viroids
   b. Antibiotics
   c. Chemicals
   d. Resistant plants
143. ELISA is the popular name of
   a. Disease
   b. Scientists (female)
   c. Serological technique
   d. Chemical
144. Tulipomania was a
   a. Name of Tulip
   b. A craze
   c. Name of Dutch Scientist
   d. Name of a city
145. Mollicute are surrounded by
   a. Cell wall
   b. No Wall
   c. Membrane
   d. Polysaccharides
146. Virus particles are measured in
   a. Nanometer
   b. Picometer
   c. Micron
   d. Cannot be measured
147. Virus are categorized as
   a. Living
   b. Non-living
   c. Between living and non-living
   d. No status as yet.

148. Plant Cuticle is specifically a barrier against
    a. All pathogens
    b. Viruses
    c. Bacteria
    d. Fungi
149. Plant Galls are produced by
    a. Viruses
    b. Host ribosomes
    c. Capsid protein
    d. Nematodes
150. Cotton leaf curl disease is caused by
    a. Specific virus
    b. Group of begomoviruses
    c. No viruses
    d. Viroids
151. Viruses multiply by
    a. Binary fission
    b. Auto catalytic reactions
    c. Sexually
    d. Asexually
152. Bird flu is a dangerous and contagious disease, because it is/has
    a. Food -borne
    b. Water -borne
    c. Many recombinants
    d. Mutates quickly
153. Tobacco mosaic virus can survive for many years because
    a. It was the first virus to be discovered
    b. Highly stable to all environments
    c. It has high thermal inactivation point.
    d. Resistant virus
154. Interveinal chlorosis in leaves is due to deficiency of
    a. N
    b. Fe
    c. Mg
    d. Cl
155. Plant Viruses can infect many forms of life
    a. Animals
    b. Man
    c. Plant kingdom
    d. All form of life
156. An agent or insect which is actively involved in the transmission of a viral disease is called
    a. Activator
    b. Carrier
    c. Vector
    d. None
157. Ratoon stunting disease of sugarcane, is named so because is named so because
    a. It infects ratooned sugarcane plants
    b. The pathogen was first found in ratooned crop
    c. The pathogen survives only in ratooned crop
    d. Ratooned plants are weaker than the normal plants.
158. Serology/ serological detection of plants viruses is a function of
    a. Non-living
    b. Nucleoprotein
    c. Antigen and antibody compatibility
    d. Dissemination of disease

159. Disease resistance (genetically) is in the
   a. Pathogen b. Host
   c. Pathogen and host d. none
160. Father of Plant Virology
   a. Beijerinck b. Bowden
   c. Watson d. Mayer
161. Plant Viruses can be dissociated and reconstituted
   a. Yes b. No
   c. Stable viruses d. Stable but mono-component viruses
162. Plant virus diseases are now more important and only next to
   a. Nematode diseases b. Bacterial diseases
   c. Fungal diseases d. Phytoplasmal diseases
163. Symptomatology is not a good criterion for the identification of virus disease because
   a. Not influenced by environment b. Influenced by mixed infection
   c. Similarity of viral symptoms d. Host reaction
164. Downy mildew are named so because they
   a. Develop on the lower/down side of leaves
   b. Appear only on the down leaves of plants
   c. Develop a downy mass on the underside of leaf
   d. Appear down on the roots.
165. Powdery Mildew of mango over winters in
   a. Dead twigs on the affected tree b. Diseased debris
   c. Alternate hosts d. Living twigs /tissue of mango.
166. Virulence of pathogen is determined by
   a. Genes of host b. Gene of the pathogen
   c. Environment d. Combination of above factors
167. Epidemiology deals with
   a. Only Spread of disease b. Influence of environmental factors
   c. None of the above d. Only Pathogens
168. Vertical Resistance is conditioned by
   a. Multiple gene b. Single gene
   c. Specific gene d. Not governed by genes

169. Peach leaf curl disease is caused by
  a. Virus b. Fungus
  c. Insect d. Nematodes

170. Plant pathology is the study of
  a. Bacteria b. Fungi
  c. Nematodes d. Plant diseases

171 Nematodes differ from insects
  a. No difference b. Having segmentation
  c. Having wing d. Having legs

172 Pest includes
  a. Only insects b. Only weeds
  c. Only pathogens d. Above all

173 Pesticide include
  a. Only fungicides b. Only nematicides
  c. Only bactericide d. Above all

174 Disease is same as
  a. Pathogen b. Insect
  c. Nematode d. Different from above

175 Virus is
  a. Nucleus acid b. Coat protein
  c. Nucleoprotein d. Small entity

176 Viruses are
  a. Obligate parasite b. Saprophyte
  c. Facultative parasite d. Facultative saprophyte

177. All nematodes remain during life
  a. Vermiform b. Pear Shaped
  c. lemon shaped d. variable in shape

178. Smuts are
  a Obligate parasites b Saprophytes
  c Facultative parasite d Facultative saprophyte

179. Rusts are
  a. Obligate parasite b. Saprophyte
  c. Facultative parasite d. Facultative saprophytes

180. Mildews are
  a. Obligate parasite b. Saprophyte
  c. Facultative parasite d. Facultative saprophytes

181. Ear cockle nematode was discovered by
  a. Smith 1673 b. Needham 1945
  c. Needam 1743 d. Sassar 1860

182. Rice brown spot caused Famine in
   a. 1945 in Bengal
   b. 1943 in Bengal
   c. 1745 in Bengal
   d. 1645 in Bengal
183. Late blight of potato is caused by Fungi
   a. Fungi
   b. Cool weather
   c. Hot weather
   d. Wrath of gods
184. Microscope was discovered by
   a. Carl von Linne
   b. Charles Darwin
   c. Anton von Leeuwenhoek
   d. deBary
185. Late blight of potato is caused byfungi was proved by
   a. Anton deBary
   b. Robert Koch
   c. Louis Pasteur
   d. Prevost
186. Virus differs from viroid because of
   a. Virus is naked but viroid had coat protien
   b. Virus has a coat protein but viroid is naked
   c. Virus is filterable but viroid not
   d. Virus is not filterable but viroid is
187. All Viruses are of
   a. lining lacks mesoderm.
   b. Becomes the intestine lumen as the nematode matures
   c. Form by secondary invagination of the gastercoel.
   d. Develops from the embryonic blastocoel.
188. Most of the plant parasitic nematodes feed on
   a. Stem
   b. Leaves
   c. Roots
   d. Flowers
189. Most zoologists believe that nematode psedocoelm
   a. Is lined by ectoderm and endoderm but the lining lacks mesoderm.
   b. Becomes the intestine lumen as the nematode matures.
   c. Form by secondary invagination of the gastercoel.
   d. Develops from the embryonic blastocoel.
193 An order which typically have cylindrical esophagus is
   a. Spirurida
   b. Dorylaimida
   c. Mononchida
   d. Enoplida
194 The perineal pattern of *Meloidogyne javanica* can be readily distinguished from *M. incognita* on the basis of
   a. Minute punctuations in tail region of M. javanica which are absent in M. incognita.
   b. "Wings" which project laterally on M. incognita but do not occur in M. javanica.

c. Striations are zig-zag in M incognita but tend to be straight and smooth in M. javanica.
d. The lateral lines which are distinct and typically double in M. javanica, but are formed by broken stria and are less distinct in *M. incognita*. distinct in M. incognita.

195. A pattern formed by transverse markings which cross the lateral field is best described as
   a. checkered b. Tessellate
   c. Areolated d. Cervical alae

196. One difference between spines and setae is that
   a. Spines tend to be larger than setae.
   b. Spines are innervated sensory structures whereas setae are simply elongate extensions of cuticle.
   c. Spines are the same as "apparent segmentation
   d. spines are typically longer and more robust than setae

197. Which of the followings has a round tail?
   a. Rhabditis b. Neotylenchus
   c. Diplogaster d. Hoplolaimus

198. *Steinernema carpocapsaeis* used against which Rice insect.
   a. Rice weevil b. Riche jassids
   c. Rice aphids d. Rice hoppers.

199. A species of nematode has two gonads in the female which both extend anteriorly. It would be correct to say
   a. Diorchic b. Prodelphic
   c. Amphidelphic d. Opistodelphic

200. The ability to distinguish between primitive and derived character states is most critical to which of the following school
   a. Cladistic b. Phenetic
   c. Evolutionary d. Fossil

201. A nematode species with an approximately equal number of males and female in given population probably reproduce by
   a. Hermaphroditism b. Meiotic parthenogenesis
   c. Mitotic parthenogenesis d. Amphimixis

202. Sex reversal may result in a male of Meloidogyne spp. with two gonads occurs when
   a. The second stage juvenile is unable to find a suitable host for a prolonged period.
   b. The individual had multiple X chromosomes, but only one Y chromosome.
   c. The environment is favorable through the period of the second molt.
   d. The process of sex differentiation is disrupted at early stage

203. Mechanoreceptors can often be recognized morphologically because
   a. They typically include openings to the anterior and a dense material occurs betweentubules of the receptors.
   b. They always occur as papillaein slight depression.
   c. They typically do not open tothe exterior.
   d. None of the above.

204. A sympathetic nervous system is
   a. One which response to pain and injury
   b. That part of the nervous system which forms a distinct principal concentrationof nerve bundles or ganglia and their interconnections.
   c. That part of the nervous system which lies primarily in the interchordal hypodermis.
   d. d.Asystem whichinnervatesspecificregions(such as viscera or the esophagus) and is generally semi- independent of the central nervous system.

205. The excretory system in adenophoreais characterized by
   a Two tubes (canals), a gland, and a single cuticle-lined duct.
   b One tube (canal) and a greatly reduced gland.
   c gland and duct
   d Pronounced duct without agland.

206. A discrete group of perikaryons canbe interpreted as a
   a. Multinucleate excretorysystem. b. Ganglion
   c. Nerve bundle d. Nerve ring

207. A type of free ending which monitor pressure, posture or mechanical deformity is a
   a. Proprioreceptor b. Dierid
   c. Commissure d. Amphid

208. Sister species are best described as species which
   a. Are so closely related that they cannot be morphologically separated
   b. Share common ancestor.
   c. Develop separately due to physical separation, such mountains,, canyons, or steams.
   d. Come from the same area but they occur on two different hosts.

209. A Peltamigrtus male can be readily distinguished from Aorolaimus male, even if the lateral field and phasmids are obscure because
   a. The stylet tubes are "tulip shaped" in male of Peltamigrtus and round in Aorolaimus.
   b. The caudal alae are adanal in Peltamigrtus and abanal in Aorolaimus.
   c. Males are extremely rare in Aorolaimu and common in Peltamigrtus.
   d. Caudal alae are "indented" in Peltamigrtus and not indented in Aorolaimus

210. Straight ovary in many plant parasitic nematode is known as
   a. Stretched ovary. b. Convoluted ovary.
   c. Curved ovary. d. None .

211 Which of the following depicts the position of antheridium in Penicillium in connection with the ascogonium?
   (a) Approaches the ascogonium only at its tip.
   (b) Rows besides the ascogonium
   (c) Remains at the base of the ascogonium,
   (d) Coils loosely around the ascogonium

212. Why zygospores are not generally formed in a culture of Rhizopus developed from a single spore ?
   (a) Due to lack of light (b) Due to shortage of oxygen
   (c) Due to absence of plus (+) and (-) strains of mycelium (d) None of these

213. Which one of the following genera does not have two ovaries?
   a. Criconemoides b. Hirschmanniella
   c. Rotylenchulus d. Heterodera

214. One important difference between Longidorus and Xiphinema is that
   a. The odentostyle extension lacks flanges in Longidorus.
   b. The amphid is split like in Longidorus and pore like in Xiphinema.
   c. Males of Longidorus typically had two testes.
   d. Males of Xiphinema have caudal alae.

215. One way in which Hoplolaimidae differs from Tylenchoridae is that
   a. In Hoplolaimidae tail is generally shorter than in Tylenchoridae.
   b. In Hoplolaimidae, the lip region is not smooth, whereas in Tylenchoridae it is striated.
   c. None of the above
   d. All of the above.

216. A juvenile of *Radpholus similis* differs from *Pratylenchus* spp. is that
   a. The tail is less tapering in Radopholus.
   b. The esophageal gland lobe overlaps ventrally in Pratylenchus.
   c. The head is more off set in Radopholus.
   d. Cuticle annulation is finer in Pratylenchus.

217. Which of the following is a good example of heterothallism ?
   (a) Spirogyra (b) Pinus
   (c) Rhizopus (d) Castor bean

218. Which one of the following has relatively minute phasmid?
   a. Hoplolaimus female b. Petsmigrtus female
   c. Tylenchulus female d. Cacopaurus female

219. Recent studies on *Caenorhabtis elegans* suggest that earlier concepts of nematode were gererally incorrect with respect to

a. Triploblastic develpoment
b. Cell consistency
c. Four moths in nematode
d. Hydrostatic skeleton

220. A large female nematode with a small stylet is recovered from an above ground plant gall. It has one ovary with numerous oocytes. This nematode is mostly likely to be

a. Heterodera
b. Nacobbus
c. Anguina
d. Ditylenchus

221. A nematode in which only lateral lines are crossed by annulations is best described as

a. Tessellate
b. Ridged
c. Aeolated
d. Striated

222. Inner label papillae (sensillae) areprobably chemoreceptor because

a. Their inner placement allows to come in contact with chemicals.
b. They are often modified to setae or bristle
c. They have openings to the external environment.
d. They have close association with amphids.

223. A fundamental difference between an odentostyle and stomatostyle is thought to be that

a. The odontostyle is a modified tooth whereas the stomatostyle is formed from the walls of the stoma.
b. The stomatostyle lumen is formed from a groove, whereas the odentostyle lumen is derived from the stma.
c. An odentostyle is developed from mesorhabdions whereasa stomatostyle is primarily developed from prorhabdions.
d. A stomatostyle is primarilydeveloped mesorhabdions whereas a n

224. An example of a nematode with a post-vulvur utrine sac is

a. Helicotylenchus
b. Pratylenchus
c. Meloidogyne
d. Anguinidae

225. A nematode with two ovaries directedin opposite direction is

a. Diorchic
b. Opistodelphic
c. Prodelphic
d. Monorchic

226. A nematode with one ovary directedanterior to vulva is

a. Diorchic
b. Opistodelphic
c. Prodelphic
d. Monorchic

227. Telogonic ovaries are because

a. Oogonia produce from germ cell at the tip of ovary
b. Oogonia produce from entire length of ovary
c. Oogonia produce from middle of ovary
d. Oogonia produce from the posterior end of ovary

228. Hologonic ovaries are because
    a. Oogonia produce from germ cell at the tip of ovary
    b. Oogonia produce from entire length of ovary
    c. Oogonia produce from middle of ovary
    d. Ogonia produce from the posterior end of ovary
229. A full complement of sensillae in thehead region of nematode, excluding amphids is
    a. 16 b. 14
    c. 12 d. 10
230. The feeding habit of root knot nematode is
    a. Ectoparasitc andMigratory b. Ectoparasitc andSedentary
    c. Endoparasitic andsedentary d. Endoparasitic andmigratory
231. Feeding habit of *Prtylenchus* is
    a. Ectoparasitc and Migratory b. Ectoparasitc and Sedentary
    c. Endoparasitic and sedentary d. Endoparasitic and migratory
232. Feeding habit of *Xiphinema* is
    a. Ectoparasitc and Migratory b. Ectoparasitc and Sedentary
    c. Endoparasitic and sedentary d. Endoparasitic and migratory
233. Disease triangle comprised of
    a. Host, pathogen andenvironment b. Resistant host, pathogen and environment
    c. Host, pathogen andtemperature d. Host, pathogen and humidity
234. Penetration of plant tissues by bacteria is mostly through
    a. Direct b. Indirect
    c. Through all above d. Wounds
235. Plant nematology is the study of
    a. free living nematodes b. plant parasitic nematodes
    c. human nematodes d. all of the above
236 Infection is a process involving
    a. Establishment of contact with susceptible tissues of the host and procurement of nutrients
    b. Establish contact with resistant tissues of the host and procure nutrients
    c. Establish contact with nonhost tissues of the host and procure nutrients
    d. Establish contact with susceptible tissues of the hostand cannot procure nutrients
237. Horizontal resistance is effective against
    a. Few races of a pathogen b. All races of pathogens
    c. Four race of a pathogen d. Two race of a pathogen

238 Vertical resistance is effectiveagainst

a. Few races of a pathogen
b. All races of pathogens
d. Four race of a pathogen
e) Two race of a pathogen

239. Hyperplasia is best described as

a. Abnormal plant growthdue to increased cell division
b. Normal plant growth dueto increased cell division
c. Abnormal plant growthdue to abnormal cell enlargement
d. Does not fit to the allabove

240. Hypertrophy is best described as

a. Abnormal plant growth due to increasedcell division
b. Abnormal plant growth due toabnormal cell enlargement
c. Normal plant growth due to increasedcell division
d. Does not fit to the all above

241. Latent infection is defined as

a. Host is infected by a pathogen and shows symptoms
b. Host gets infected by a pathogen and is symptomless
c. Host is not infected by pathogen and shows symptoms
d. Host is infected by pathogen and develop mild symptoms

242. Damping-off is best described as

a. Destruction of seedling near soil line by pathogen and falling on ground
b. Destruction of seedlings by pathogen at the middle of shoot and Falling on ground
c. Destruction seedlings bywind and falling on ground
d. Destruction of seedling byhailstorm and falling on ground

243. Dikaryotic mycelium is because

a. Mycelium contains two sexually compatible nuclei per
b. Mycelium contains two sexually incompatible nuclei per cell
c. Mycelium contains two asexually incompatible nuclei per cell
d. Mycelium contains two asexually compatible nuclei per cell

244. Dieback is known because of

a. Progressive death of plant parts starting from the tip
b. Progressive death of plant parts starting from the base
c. Progressive death of plant parts starting from insect attack
d. Progressive death of plant parts starting from nematode infection

245. Density gradient centrifugation is because

a. Particles are separated by in layer according to their density
b. Particles are separated by in layer according to their molecular weight
c. Particles are separated by in layer according to their molecular weight and density
d. Particles are separated by in layer according to the speed of centrifuge

246. Cytoplasmic resistant is because
  a. Resistance is controlled by genetic material present in cytoplasm
  b. b.Resistance is controlled by genetic material present in golgi body
  c. Resistance is controlled by chromosome
  d. Resistance is not controlled by genetic material

247. Cross protection is a phenomenon in which
  a. Prior infection by aggressive strain of a pathogen protects the plant against non aggressive pathogen
  b. Prior infection by non aggressive stain of a pathogen protects the plant against aggressive pathogen
  c. Prior infection by non aggressive stain of a pathogen does not protect the plant against challenged by aggressive pathogen
  d. Prior infection by aggressive stain ofa pathogen does not protect the plant against challenged nonaggressive pathogen

248. Disinfestant differs from disinfectant because
  a. Disinfestant does not eradicate the infection whereas disinfectant kills infestation
  b. Disinfectants are used to kill germs, viruses, and bacteria on non-living surfaces, while disinfestants are used to eliminate insects and other pests from an environment.
  c. Disinfestant eradicate both infection and infection whereas disinfectant does not.
  d. They do not differ

249. Forma specialis (F. sp.) is a categoryof pathogen which infects
  a. Large number of host plant
  b. Host plants within a genus or species
  c. Infect host plant
  d. Does not infect

250. Fission is a process best described as
  a. Transverse splitting of bacterial cell in two
  b. Transverse splitting of viral cell in two
  c. Transverse splitting of fungal cell in two
  d. All above

251. Phytoanticipins
  a. Inhibitory antimicrobial compounds present in plant cell before infection
  b. Inhibitory antimicrobial compounds produce in plant cell in response to infection
  c. Noninhibitory Beneficial compoundspresent in plant cell before infection
  d. Inhibitory antimicrobial compounds present in cell of pathogen

252. Phytoalexins are defined as
   a. Substance which is present in the plant but are not inhibitory to development of a pathogen
   b. Substance which is present in plant and are inhibitory to development of a pathogen
   c. Substance which is present not in plant but are produced by the pathogen
   d. Substance which is present in plant but encouragea the development of the pathogen

253. Resistant plants differ from tolerant plant
   a. Resistance is not genetically controlled but tolerance is genetically controlled
   b. Resistance is genetically controlled but tolerance is not geneticallycontrolled
   c. Resistance is not genetically controlled but tolerance is also not genetically controlled
   d. Do not differ

254. Perithecium, a fungal fruiting is best described as
   a. An open cup shaped ascocarp b. Flask shaped ascocarp
   c. A closed cup shaped ascocarp d. Does not fit above

255. Apothecium, a fungal fruiting is best described as
   a. An open cup shaped ascocarp b. Flask shaped ascocarp
   c. A closed cup shaped ascocarp d. Does not fit above

256. Plant pathogenesis-related protein are
   a. Induced in plant in response to infection and are mostly toxic
   b. Induced in plant in response to infection but not toxic
   c. Not induced in plant in response to infection but are already present and are not toxic
   d. Not induced in plant response to infection but are already present and are toxic

257. Juvenile differs from adult nematode because
   a. Juvenile has gonads whereas adult lacks
   b. Juvenile is smaller in size whereas adult is bigger in size
   c. Juvenile lacks gonads whereas adult has
   d. They do not differ

258. Number of molts in nematode is
   a. 2 b. 3
   c. 4 d. 5

259 J2 hatch from eggs of most of the plant parasitic nematodes except
   a. *Xiphinema* *b.* Meloidogyne
   c. *Pratylenchus* d. *Trichodorus*

260. Root knot nematode differs from cystnematode because
   a. Root knot nematode induces cyst on roots whereas cyst nematode galls.
   b. Root knot nematode induces gall on roots whereas cyst nematode also galls.
   c. Root knot nematode induces gall on roots whereas cyst nematode cysts.
   d. Root knot nematode induces cysts on roots whereas cyst nematode also cysts.
261. Root knot nematode differs from cyst nematode because
   a. Root knot nematode induces joint cells in roots whereas cyst nematode syncytium.
   b. Root knot nematode induces syncytium.in roots whereas cyst nematode joint cells.
   c. Root knot nematode induces syncytium in roots whereas cystnematode also syncytium.
   d. Root knot nematode induces joint cells in roots whereas cyst nematode also joint cells.
262. One nematode genera that is virus vector
   a. *Pratylenchus*   b. *Tylenchus*
   c. *Trichdorus*   d. *Paratylencus*
263. Grape leaf fan virus is transmitted by
   a. *Xiphinema americanum*   b. *Xiphinema index*
   c. *Xiphinema ifacolum*   d. *Xiphinema insigne*
264. Tobacco Rattle viruses (Tobra. arevectored by genus
   a. *Xiphinema*   b. *Longidorus*
   c. *Paratrichodorus*   d. Meloidogyne
265. Nematode-transmitted polyhedralviruses (NEPO) are vectored by genus
   a. Longidorus   b. Trichodoru
   c. Paratrichodorus   d. Pratylenchus
266. Gelatinous mass in root knot nematode is produced from
   a. Vulva   b. Anus
   c. Rectal glands   d. Excretory pore
267. Root knot nematode was firstreported by
   a Van Gundy (USA.   b Golden (UK)
   c Berkeley (UK)   d McKenry (USA.
268. Recently National Nematology lab inUAF has been established by
   a. Dr. Riaz Ahmad   b. Dr. Safdar A. Anwar
   c. Dr. Nazir Javed   d. Dr. Sajid Aleem
269 First course of nematology course was taught at University of Agriculture, Faisalabad, Pakistan by
   a. Safdar Anwar   b. Inam Ullah. Khan
   c. A G. Kausar   d. R. A Chohan

270. Elicitors are
   a. Molecules produced by pathogens toinitiate disease
   b. Molecules produced by host to defendagainst invasion
   c. Molecules are produced by pathogen to protect the host
   d. Not fit to above.

271. Stubborn disease of citrus is caused by
   a. Spiroplasma
   b. Mycoplasma
   c. L-form bacteria
   d. Xylem limited fastidious bacteria

272. The cell wall is surrounded in some bacteria by a polysaccharide in the form of a definite layer called
   a. Plasma membrane
   b. Capsule
   c. Outer membrane
   d. Basal body

273. The minute hair-like superficial appendages, covering the surface of most ofthe Gram –ve bacteria are called
   a. Flagella
   b. Capsule
   c. Pilli
   d. None of the above

274. The photosynthetic bacteria have
   a. Mesosomes
   b. Ribosomes
   c. Lamellae
   d. Flagella

275. The particles concerned with bacterial genome replication and septum formation during the cell wall division arecalled
   a. Mesosomes
   b. Thylakoids
   c. Plasmids
   d. Ribosomes

276. Mucopeptide forms major portion(85%) of the cell wall in the.
   a. Gram –ve
   b. Fastidious
   c. Gram +ve
   d. L-form

277. The most important factor in the multiplication of plant pathogenic bacteria is
   a. Temperature
   b. Moisture
   c. Relative humidity
   d. Rainfall

278. Plant pathogenic bacteria are
   a. Saprophytes
   b. Parasites
   c. Facultative parasites
   d. All of the above

279. Xylem-limited fastidious bacteria be cultured on
   a. PDA
   b. Nutrient agar
   c. Axenic culture
   d. None of the above

280 The nematodes lack one of the systems that is
   a. Digestive system
   b. Reproductive system
   c. Nervous system
   d. Circulatory system.

281. Mycophagous are

a. Feeding on fungus
b. Feeding on bacterium
c. Feeding on dead matter
d. Feeding on plants

282 People who love and have knowledge of mushrooms are called

a. mycophiles
b. mycophobes
c. Mushroom growers
d. Mushroom hunters

283. An association of roots with fungus is called

a. mycorrihza
b. Saprophytic
c. commensalism
d. None

284. How many phases are in composting?

a. Two
a. Three
b. Four
c. Five

285. The branch of biology dealing with the functions and vital processes of living organisms and their components is called

a. Physiology
b. Morphology
c. Histology
d. Microbiology

286. A tissue covering mushrooms as they develop is called

a. Veil
b. Volva
c. Ultra structure
d. Cap

287. Chinese mushroom is also called

a. Shiitake mushroom
b. oyster mushroom
c. Button mushroom
d. black morels

288. Mushroom is Enriched in

a. Protein
b. Carbohydrates
c. Sugar
d. Starch

289. Mushroom contains how much water contents

a. 50%
b. 60%
c. 90%
d. 30%

290. *Ganoderma lucidum* is a leader of

a. Medicinal mushroom
b. Poisonous mushroom
c. Cultivated mushroom
d. Non cultivated mushroom

291. The haploid mycelium derived from basidiospores is called

a. Spawn
b. Primary mycelium
c. Inoculum
d. Culture

292. A culture that contains a population of only one specie is called

a. Pure culture
b. Spawn
c. Pulp
d. Secondary metabolism

293. The annulus is a

a. Ring
b. Spoil
c. Lesion
d. Patch

294. Saprophagous

a. Feeding on decaying matter
b. Feeding on living matter
c. Feeding on both living and nonliving matter
d. Prepare their own food

295. A technique of mushroom cultivation in which plants and fungus are grown together to obtain fruiting bodies is called

a. Semi cultivation
b. Cultivation
c. Farming
d. Outdoor cultivation

296 A dot like visible indication of mushroom primodium formation in a mycelium that can be distinguished and will form mushroom is called

a. Pinhead
b. Spawn
c. Stalk
d. Cap

297. The occurrence of more than one distinct form in the life cycle is called

a. Pleomorphic
b. Plasmogamy
c. Polygenic
d. polyploidy

298. Any of large group of mostly tough, wood inhibiting mushrooms, which bear their spores, is called

a. Protroph
b. Psycotrophic
c. Pyrethrins
d. Polypores

299. The mouths of the tubes in boletes and polypores are called

a. Openings
b. Poses
c. Primordium
d. Spore

300. The substrate in which mushroom mycelium has developed and which will be used as a seed in propagation for mushroom production is called

a. Compost
b. Seed bed
c. Spawn
d. Seed

301 The planting of mushroom spawn in to the prepared substrate or compost is called

a. Mushroom cultivation
b. Spawning
c. Spwan running
d. Inoculation

302. Culture derived by isolation of a single spore and its subsequent germination and development into mycelium is called

a. Pure culture
b. Spawn
c. Mixed culture
d. Spore culture

303 Microscopic cell capable of developing into an adult organism

a. Seed
b. Embryo
c. Spore
d. Fruit

304. The scientific term for the stem of the mushroom is called

a. Stipe
c. Pileus
c. Volva
d. Annulus

305. A layer of hyphae in the central part of a gill is called

a. Trama | b. Fiber
c. Developing tissues | d. Veil

306. The knob in the center of the pileus of certain mushroom is called

a. Ultrastruture | b. Universal veil
c. Umbo | d. Vacoule

307 *Amanita phalloides* is the group of mushroom, which is

a. Poisonous | b. Edible
b. Protein enriched | d. Rich in vitamins

308. Poisonous mushroom are mostly

a. Bright in color | b. Dull in color
c. Colorless | d Yellowish

309. Pathology is derived from two words of

a. English | b. Greek
c. Latin | d. Roman

310 The five kingdoms classification of fungi proposed in 1969 by

a. Linnaeus | b. Persoon
c. Bessey | d. Whittaker

311. The DNA based classification of fungi is proposed in 1996, separated the lower fungi under the kingdom

a. Stramenopila | b. Protista
c. Fungi | d. Monera

312. On the global basis over all annual crop losses is about

a. 38% | b. 40%
c. 31% | d. 35%

313. Annual crop losses due to different diseases is abou

a. 13% | b. 11%
c. 10% | d. 15%

314. The famous Bengal famine due to brown leaf spot of rice occurred during the year

a. 1940 | b. 1939
c. 1941 | d. 1943

315. Yellowing of normally green tissue due to chlorophyll degradation is known as

a. Necrosis | b. Mosaic
c. Chlorosis | d. Scorching

316. In fungi the homothallic sexual reproduction takes place on

a. Single thallus | b. Compatible thallus
c. Multi thallus | d. None of these

317. *Asprgillus niger* is also known as

a. Sooty mold | b. Bread mold
c. Black mold | d. All of them

318. Production of Zoospores is characterized to class

a. Ascomycetes b. Chytridiomycetes

c. Basidiomycetes d. Myxomycetes

319. The bacterial cells that lack flagella are known as

a. Atrichous b. Monotrichous

c. Amphitrichous d. All of them

320. First Quarantine law was implemented in 1960 by

a. English b. American

c. Indian d. French

321 In standard Bordeaux mixture the ratio of different contents is

a. 2:2:50 b. 2:2:100

c. 1:8:45 d. 4:4:50

322Abnormal enlargement of plant tissues due to some diseases is called

a. Hypersensitive reaction b. Hyperplasia

c. Hypertrophy d. Both b & c

323. The term witches' broom refers to

a.. Stunting of whole plant b. Smalling of leaves

c. Dense clustering of branches d. Tree bark shredding

324. The first idea of mycotoxin came from England due to Turkey"X" disease in the year

a. 1960 b.. 1964

c. 1966 d. 1968

325. Aflatoxin is produced by

a. *Penicillium notatum* b. *Penicillium rubrum*

c. *Asprgillus niger* d. *Asprgillus flavus*

326. The ever first evaluated fungicide based

a. Copper b. Sulphur

c. Organophasphate d. None of these

327. Mostly Black or stem rust of wheat occurs in the month of

a. December b. January

c. February d. March

328. Gram blight is a serious disease of

a. Hot regions b. Humid regions

c. Dry regions d. Cooler region

329. Citrus canker disease is caused by bacterium of

a. *Erwinia* spp. b. *Pseudomonas* spp.

c. *Xanthomonas* spp. d. None of these

330. Late blight of potato caused famine in Ireland in the year

a. 1940 b. 1943

c. 1845 d. 1841

331. In case of apple scab disease development favours by

a. Cool wet weather b. Warm dry weather

c. Cool dry weather d. Warm humid weather

332. Alternate host of brown rust of wheat is

a. *Puccinia recondita* b. *Berberris vulgaris*

c. *Thallictrum flavum* d. None of these

333. Foot rot disease of rice is caused by a

a. Nematode b. Fungi

c. High pH of water d. Soil bacterium

334. Teliospores are produced in a fungal

a. Mushrooms group called b. Rusts

c. Smuts d. B&c

335. The fungi which absorbed their foodfrom wood called as

a. Folicolous b. Fungicolous

c. Lignicolous d. Coprophilous

336. Amanita mushrooms included ingroup of

a. Edible b. Poisonous

c. Domestic d. Cultivated

337. Bacteria reproduced asexually by means of

a. Binary fission b. Budding

c. Fragmentation d. None of these

339. Mycoplasmas are prokaryotes which lack

a. DNA b. RNA

c. Cell membrane d. Cell wall

340. Needham observed nematodes within small rounded wheat kernels for the first time in

a. 1741 b. 1743

c. 1745 d. 1747

341 Koch proposed “Koch’s Postulates “which are used to confirm the pathogenicity in the year.

a. 1880 b. 1882

c. 1782 d. 1780

342. In 1886 'Tobacco Mosaic virus'disease is firstly reported by

a. Ivanowski
b. Beijerinck
c. E. F. Smith
d. Mayer.

343. Up till1936, it is believed that virus actually consisted of not only proteins but also a small amount of

a. RNA
b. DNA
c. Both RNA& DNA
d. None of these

344. In 1967, Doi et al in Japan observed a new microorganism group named as

a. Viruses
b. Bacteria
c. Mollicutes
d. Nematodes

345. Control of citrus tristeza and some other viral diseases has been achieved by

a. Antibiotics
b. ELISA
c. Antagonistic fungi
d. Cross protection.

346. In the late 1980s, a new technology was introduced to control plant diseases.

a. Genetic engineering
b. Biocontrol
c. Biotechnology
d. None of these

347. Cloudy weather in spring favours the scab disease development with optimum temperature of

a. 16-24 °C.
b. 10-20 °C.
c. 24-28 °C.
d. 26-32 °C.

348. The fruiting body of *Venturia inaequalis* fungus called as

a. Perithecium
b. Appothecium
c. Pseudothecium
d. None of these

349. Early blight of potato a fungal disease caused by

a. *Phytophthora infestans*
b. *Ascochyta rabiei*
c. *Alternaria solani*
d. *Alternaria alternata*

350. The juice obtained from red rot infected canes gives bad odour like

a. Rotten fish
b. Rotten egg
c. Methane gas
d. Alcohol

351 In case of red rot of sugarcane the main source of primary inoculum is

a. Seed setts
b. Infected soil
c. Infected water
d. Infected air

352. Zn deficiency is a common problem in

a.. Wheat
b.. Rice
c. Maize
d. Sugarcane

353. Bacterial blight of cotton is alsocalled as

a. Black arm disease
b. Angular spot of cotton
c. Bacterial canker
d. Both a and b

354. Ear cockle of wheat is a nematodedisease which attack on
  a. Roots b. Stems
  c. Spike d. Leaves
355. Tirak or Bad Opening of Bolls of Cotton is mainly due to deficiency of
  a. P b. N
  c. K d. N and P
356. Potato cyst nematodes were also called as
  a. Root gall nematode b. Root lesion nematode
  c. Potato golden nematode d. Both a and b
357. Cotton Leaf Curl Virus (CLCuV) disease appeared firstly in Multan in the year
  a. 1967 b. 1969
  c. 1970 d.
359. Cotton Leaf Curl Virus introduced inPakistan through import of cotton seed from
  a. Egypt b. Russia
  c. India d. USA
360. The typical CLCuV symptoms are leaf enation and
  a.. Vein clearing b. Vein thickening
  c. Vein banding d. Vein twisting
361. In nature CLCuV virus is transmitted by
  a. Jassid b. Thrips
  c. Whitefly d. Both b and c
362. Incubation period for CLCuV virus varies from
  a. 10-15 days b. 15-20 days.
  c. 20-25 days d. 25-30 days
363. The alternate hosts of CLCuV is
  a. Eggplant b. French bean
  c. Okra d. All of these
364. Banana Bunchy Top Virus (BBTV) is transmitted through
  a. Banana Aphids b. Banana white fly
  c. Black jassids d. None of these
365. Thermal inactivation point of potato leaf roll virus is
  a. 50-60°C b. 70-80°C
  c. 80-90°C d. 90-100°C
366. The mild mosaic is the general symptom clearly visible in case of
  a. Potato Leaf Roll Virus b. Potato Poty virus A
  c. Potato Poty virus Y d. *Potato virus X*
367. Sugarcane mosaic disease was first recorded in 1892 from
  a. Java b. Japan
  c. Brazil d. USA

368. *Citrus tristeza Virus* belongs to group
    a. Begomoviridae b. Nepo virus
    c. Closteroviridae d. None of these
369. Citrus canker disease is favoured by mild 20-30 °C and
    a. Wet weather b. Dry weather
    c. High pH of H2O d. Low pH of H2o
370. Angular leaf spot of cotton mostly effect the plant parts
    a. Branch b. Bolls
    c. Leaves d. All of them
372. The wheat rust fungi produced spore during II stage of life is
    a. Aeciospores b. Teleutospores
    c. Uredospores d. Basidiospores
373. The wheat rust fungi are obligate parasites which belongs to family
    a. Uredinaceae b. Pucciniaceae
    c.. Melampsoraceae d. Ustilaginaceae
374. The heteroecious wheat rust species typically develop which fruiting bodies on another host
    a. Telia and Uredia b. Pycnia and Aecia
    c. Telia and Basidia d. Both b and c
375. In some fungi mycelium form thick strands known as
    a. Sclerotia b. Rhizomorphs
    c. Appressorium d. None of these
376. The reproductive organs arise from only a portion of the thallus called as
    a. Eucarpic b.. Holocarpic
    c. Teleomorph d. Anamorph
377. If the spore cell becomes enveloped in a thick wall before separating known as
    a. Conidia b.. Sporangiospores
    c. Zoospores d. Chlamydospores
378. Fungi producing both male and female sex organs on the same thallus called as
    a. Dioceous b. Heteroecious
    c. Hermaphrodite d. Both b and c
379. *Synchytrium,* and *Physoderma* are the members of phylum
    a. Chytridiomycota b. Myxomycota
    c. *Zygomycota.* d. Bisidiomycota
380. Pale brown spots causing target board effect on leaves, appear in
    a. Black wart b. Hollowleaves
    c. Late blight d. Early blight
381. CLCuV morphologically consist of
    a. Circular ssRNA b. Rod shape ssRNA
    c. Rod shape ssDNA d. Circular ssDNA

382 Chemically a virus particle consist of

a. 5-40% nucleic acid & 60-95% protein

b. 15-50% nucleic acid & 50-85%protein

c. 20-60% nucleic acid & 40-80%protein

d. 5-25% nucleic acid & 75-95%protein

383. In bacteria respiration take placethrough

a. Mitochondria b. None

c. Ribosomes d. Plasma membrane

384 Members of family Erysiphaceae cause serious group of diseases called

a. Downy mildew b. White rusts

c. Powdery mildew d. Papery mildew

385 Peach leaf curl was first originated in

a. Europe b. China

c. Japan d. USA

386. Spores of wheat bunt give bad smell like

a. Rotten fish b. Rotten egg

c. Alcohol d. Onion

387. How many basidiospores are produced in uredinales

a. 8 b. 4

c. 2 d. 7

388. Crown gall of tomato disease is produced by

a. *Xanthomononas axanopodi* b. *Pseudomononas syringe*

c. *Agrobacteriam tumefaciens* d. *Erwinia crotovora*

389. White rust crucifer is produced by a fungus named

a. *Plasamopara viticola* b. *Albugo candida*

c. *Ervisiphe polygony* d. *Alternaria solani*

390 Standard moisture contents for healthy seeds is

a. 18% b. 14%

c. 12% d. 08%

391. The initial seed obtain from selected individual plants of a particular variety is

a. Nucleus seed b. Foundation seed

c. Breeder seed d. Primary seed

392. The standard purity in register seed should be

a. 92% b. 93%

c. 96% d. 98%

393. The most abundant storage fungi is

a. *Ganoderma* b. *Aspergillus*

c. *Pythium* d. None

396. which one pathogen is cannot grow on culture media
   a. Bacteria b. Phytoplasma
   c. L-form bacteria d. Fungi
397. Nematodes possess all the physiological systems like animals except
   a. Digestive b. Circulatory
   c. Nervous system d. Both band c
398. Dodders are complete parasites and belong to family
   a. Loranthaceae b. Cuscutaceae
   c. Convolvulaceae d. Orobanchaceae
399 Poor root system and burning appearance of crops is due to
   a. N deficiency b. P and K deficiency
   c. K deficiency d. Zn deficiency
400. Common N symptom is
   a. Chlorosis b. Reddening
   c. Pinkish colour d. None
401. Loose and covered smut of oats infect
   a. Seedlings b. Local
   c. Blossom d. None of these
402. Soft rot of potato also called as
   a. Bacterial rots b. Grey rots
   c. Black leg d. Black arm
403. Plant disease pyramid components are
   a. 3 b. 4
   c. 5 d. 6
404. Root knot nematodes move rapidly in soil type
   a. Silt b. Sand
   c. Sandy loam d. Clay loam
405. Citrus nematodes are the main cause of
   a. Quick decline b. Slow decline
   c. Spreading decline d. Both a and b
406. Root knot nematodes can produce a complex syndrome in combination with
   a. Potex viruses b. Alkaline soils
   c. *Fusarium* spp d. *Alternaria* spp
407. The name "father of phytobacteriology" is given to
   a. Louis Pasteur b. DeBary
   c. Prevost d. E. F. Smith
408. The book "Fifty years of investigations on plant diseases" was written by
   a. A.G. Kausar b. A. Haffiz
   c. A. Sattar d. Rai Bahadur Ch.

409. The mycelium used for the commercial production of mushrooms called as

a. Mushroom seed
b. Mushroom substrate
c. Spawn
d. Spores

410. Healthy association between two organisms not damaging each other known as

a. Parasitism
b. Synergism
c. Syndrome
d. Antagonism

411. Old bunt of wheat is also called

a. Karnal bunt
b. Stinking smut
c. Partial bunt
d. Head smut

412. Mollicutes found in phloem of plants are

a. Bacteria with flagella,
b. Virus without protein coat
c. Wall less mycoplasma like bodies
d. L-form bacteria

413. Virus consists of

a. RNA and DNA
b. RNA or DNA
c. RNA and DNA and protein
d. RNA+protein or DNA+Protein

414. Biological control is the control of pathogens by

a. Manually,
b. Chemical
c. Plant extracts
d. Living organisms.

415. Hypertrophy is due to the

a. Excessive division of cell.
b. Enlargement of cells
c. Reduction in size of cells.
d. Reduction in division of cells.

416. Most of the fungi are

a. Autotrophic
b. Semi-auto-tropic
c. Heterotrophic
d. Chemotrophic

417. First Quarantine law was implemented in 1960 by

a. English
b American
c Indian
d French

418. Coelomyces is used to kill

a. Insects
b. Nematodies
c. bacteria
d. viruses

419. Fungi produce

a. Vitamins and hormones
b. Enzymes, vitamins and organic acid
c. Enzyme, organic acid and hormones,
d. Enzyme, vitamins organic acid and hormones.

420. Coenocytic fungi have which hyphae
  a. Septate b. Aseptate
  c. Transparent d. Tubulore pore cap.

421. Karyogamy is the fusion of
  a. cells b. cytoplasm
  c. Nuclei d. Hyphae

422. Smut diseases produce which colour spores
  a. White b. Yellow
  c. Black d. Brown

423. Capsid is the protein of
  a. Viroid b. Virus
  c. Bacteria d. Prions

424. Cross protection is the phenomenon of injecting
  a. Aggressive strain b. Nematode
  c. Fungi d. Milder strain

425. Ascocorp in ascomycetes is
  a. sac like mycelia b. Fruiting body having asci
  c. Ascus d. Ascospore

426. Virion is a mature infectious particle of
  a. Viroid b. virus
  c. Prions d. Capsid

427. Beijerink is considered the father of
  a. Virology b. Bacteriology
  c. Nematology d. Mycology

428. Robert Hartig is considered as father of
  a. Forest Pathology b. Microbiology
  c. Phycology d. Entomology

429. When a pathogen spreads from the initial point of entry, it produces
  a. Local symptoms b. Galls
  c. Knots d. Systemic symptom

430. Necrosis is the
  a. Change in the colour of infected plant part b. Death of cells
  c. Ring spotting d. Enation

431. The organisms involved in the transport of virus from one plant to other are called
  a. Carrier b. Transporter
  c. Vector d. Spreader

432. Geminiviruses are mostly transmitted by

a. Aphids b. Whitefly

c. Leaf hoppers d. Jassids

434. CLCu V is a geminivirus having genetic material

a. Phytobacteria a. ss RNA

b. Mollicutes b. ss DNA

435. Plant viruses are classified on the basis of

b. Fruiting body having asci a. Morphology

c. Ascus d. Ascospore acid

436. Root knot of vegetables is caused by

a. Viruses b. Bacteria

c. Nematodes d. Fungi

437. Anthridium in some fungi i s

a. Male sex organ b. Female sex organ

c. Asexuall spore d. Sexual spore

438. Cup like sexual fruiting body in ascomycetes is

a. Perthcium b. Apothecium

c. Synnemata d. Pseudothecium

439. Sexual fruiting body in which conidiophores united at the base but apart at the top

a. Pycnidium b. Acervulus

c. Synnemata d. Sporodochium

440. A sexual ascomycete spore is

a. Ascospore b. Archiascospore

c. Conidiophore d. Pyrenospore

441. Each basidium has number of basidiospores

a. 2 b. 4

c. 9 d. 12

442. Rust fungi may produce different stages

a. 3 b. 4

c. 5 d. 6

443. Members of family pucciniaceae mostly cause diseases

a. Smuts b. Rusts

c. Anthracnose d. Leaf spots

444. Biflagellate zoospores are present in class

a. Ascomycetes b. Basidiomycetes

c. Zygomycetes d. oomycetes

445. Mildews are plant diseases in which pathogen can be seen as a growth

a. Inside the host b. On the surface of the host

c. In xylem vessel d. In phloem vessel

446. Hyperplasia is due to the
   a. Enlargement of cells
   b. Increase in the number of cell
   c. Reduction in size of host
   d. Death of cells.

447. Narrow lesion on leaves or stem are called
   a. Canker
   b. Streaks or stripe
   c. Blight
   d. Necrosis

448. Permanent wilting is caused by
   a. Abiotic stresses
   b. Plugging of xylem vessel
   c. Attack of nematodes
   d. Injury

449. Damping off is plant disease, in which plants died
   a. At flowering stage
   b. At seedling stage
   c. After maturity
   d. At fruit formation

450. In green ear diseases flowers are converted into green leaf structures called
   a. Greening
   b. Malformation
   c. Phyllody
   d. Transformation

451. Amphitrichous bacteria have
   a. No flagella
   b. One flagellum at each end
   c. Single flagellum at one end.
   d. Flagella on all over the body.

452. Bacterium belongs to
   a. Eukaryotes
   b. Prokaryotes
   c. Both prokaryote and eukaryote
   d. Autotrophs

453. Psychrophilic bacteria can grow at
   a. 0°C
   b. 30°C
   c. 50°C
   d. 93°C

454. Is Immunity developed during life time
   a. Acquired immunity
   b. Artificial immunity
   c. Deficiency syndrome
   d. All of the above

455. Aerobic are those organisms which live in the presence of
   a. CO2
   b. O2
   c. N2
   d. H2

456 Alternate host is that host which bears the pathogen
   a. In the major host crop
   b. In the absence of major host crop
   c. Host of more than two types of pathogen
   d. Collateral host

457. Phenomenon in which one organism inhibits or kill another is called
   a. Parasitism
   b. Saprophyte
   c. Antagonism
   d. Anastomosis

458. A protein produced in warm blooded animals in reaction to an injected foreign protein is
   a. Antigen b. Antibody
   c. Epitope d. Antiserum
459. Ascospore are produced in
   a. Pycnidium b. Ascus
   c. Synnemata d. Basidium
460 A chemical which prevents the multiplication of bacteria without killing then
   a. Bactericide b. Bacteriocin
   c. Bacteriostatic d. Bacteriophage
461. Flora and fauna of a region is
   a. Biotroph b. Biotype
   c. Bioassay d. Biota
462. Rapid browning of leaves, flowers and stem is
   a. Blister b. Blotch
   c. Blight d. Necrosis
463. A disease that is permanently present in a locality in mild or severe form.
   a. Epidemic b. Endemic
   c. Pandemic d. Epidemic rate
464. A compound that kills fungi
   a. Fungistatic b. Fungistasis
   c. Fungicide d. Fumigant
465. Control of a disease by preventing its introduction into disease free location is
   a. Eradication b. Exclusion
   c. Exotic d. Protectants.
466. Plants that react to certain viruses with specific symptoms.
   a. Indicator plants b. Host plants
   c. Alternate hosts d. Colleteral hosts.
467. A chain of ribonucleotides that codes for a specific protein
   a. mRNA b. tRNA
   c. rRNA d. RNA
468. Nanometer is equal to
   a. 1/10 microns b. 1/100 microns,
   c. 1/1000 microns d. 1/10000 microns
469. Combination of a sugar and a base molecule in nucleic acid is
   a. Nucleotide b. Nucleoside
   c. Nucleoprotein d. Nucleocapsid.
470. Micron is equal to
   a. 1/100 mm b. 1/1000 mm
   c. 1/10000 mm d. 1/10mm

471. The state in which host is infected but does not show symptoms

a. Masked symptom b. Latent infection
c. Latent period d. incubation

472. Bacteria that temporarily or permanent lost their ability to produce a cell wall due to exposure to antibiotic

a. L-form bacteria b. Phytoplasma
c. Spiroplasm d. MLD's

473 A parasite that can grow and multiply only on living organisms

a. Parasite b. Obligate parasite
c. Saprophyte d. Facultative saprophyte

474 A pore like opening in perithecia and pycnidia through which spores are released

a. Vesicle b. Hymenium,
c. Locule d. Ostiole

476. The ability of a pathogen to cause disease is

a. Pathogensis b. Pathogenicity
c. Pathovar d. Parasexualism

477. The external visible appearance of an organisms

a. Genotype b. Biotype
c. Phenotype d. Dimorph

478. An enzyme that break down RNA

a. Ribonuclease b. Polymerase
c. isomerase d. None of the above

479. Fungi in which no form of spore has been seen

a. Homothalic, b. Hetero thallic
c. Sterile fungi d. Hermaphrodite

480. Copying of mRNA into proteins

a. Transformation b. Translocation
c. Transmission d. Translation

481. A sub-viral circular RNA component of some RNA viruses

a. Viroid b. Virion
c. Virusoid d. Prions

482. Removal of food by a pathogen from its host is call

a. Symbiosis b. Parasitism
c. Pathogensis d. Antagonism

483. One unit of inoculum of any pathogen is

a. Propagule b. Fragment
c. Infectant d. Infestant

484. Bacteria penetrate plant surface by
   a. Natural openings b. Wound
   c. Hydathodes d. All of the above
485. Cutin is degraded by
   a. Cellulase b. Cutinase
   c. Amylase d. Pectinase
486. Tylosis is formed in
   a. Phloem vessels b. Xylem vessels
   c. Roots d. Leaves
487. Rod-shaped bacterium is
   a. Cocci b. Bacillus
   c. Spiral d. Polyhedral
488. Cell division of plants is regulated by
   a. Auxins b. IAA
   c. Cytokinins d. Gibberllins
489. Death of shoots, branches and roots starting at the tip is called
   a. Dieback b. Canker
   c. Scab d. Blotch
490. The determination and study of the cause of the disease is
   a. Biology b. Etiology
   c. Epidemiology d. Epiphytotic
491. Hyaline stands for
   a. Brown b. Black
   c. Colourless d. Blue
492 Production of Zoospores is characterized to class
   a. Oomycetes b. Chytridiomycetes
   c. Basidiomycetes d. Myxomycetes
493. In 1886 'Tobacco Mosaic virus' disease was firstly reported by
   a. Ivanowski b. Beijerinck
   c. E. F. Smith d. Mayer
494. Foot rot disease of rice is caused by
   a. Nematode b. Fungi
   c. High pH of water d. Soil bacterium
495. Control of citrus tristeza and some other viral diseases has been achieved by
   a. Antibiotics b. ELISA
   c. Antagonistic fungi d. Cross protection
496. Mycoplasmas are prokaryotes which lack
   a. DNA b. RNA
   c. Cell membrane d. Cell wall

497 Cotton Leaf Curl Virus (CLCuV) disease appeared firstly in Multan in the year

a. 1967 b. 1969

c. 1970 d. 1972

498. Incubation period for CLCuV virus varies from

a. 10-15 days. b. 15-20days.

c. 20-25 days. d. 25-30 days.

499 Banana Bunchy Top Virus (BBTV) is transmitted through

a. Black Aphids b. Banana white fly

c. Black jassids d. None of these

500. Sugarcane mosaic disease was first recorded in 1892 from

a. Japan b. Java

c. USA d. Brazil

501. Citrus Aphids transmit the tristeza virus through a manner

a. Non-persistent b. Semi-persistent

c. Persistent d. None of these

502. Pale brown spots causing target board effect on leaves, appear in

a. Black wart b. Hollowleaves

c. Late blight d. Early blight

503. The wheat rust fungi produced spore during II stage of life

a. Aeciospores b. Teleutospores

c. Uredospores d. Basidiospores

504. The members of division Gracilicutes bacteria have a cell envelope in

a. Gram+ve b. Gram-ve

c. Neutral d. Both a and b

505. *S*oft rot of potato also called as

a. Bacterial rots b. Grey rots

c. Black leg d. Black arm

506. Root knot nematodes can produce a complex syndrome in combination with

a. Potex viruses b. Alkaline soils

c. *Fusarium* spp d. *Alternaria* spp

507. *N*eedham observed nematodes within small rounded wheat kernels for the first time in

a. 1741 b. 1743

c. 1745 d. 1747

508. The ever first evaluated fungicide based on

a. Copper b. Sulphur

c. Organophosphate d. None of these

509. *Asprgillus niger* is also known as

a. Sooty mold b. Bread mold

c. Black mold d. All of them

510 The famous Bengal famine due to brown leaf spot of rice occurred during the year

a. 1939 b. 1941

c. 1943 d. 1940

511. Yellowing of green tissues due to chlorophyll degradation is known as

a. Necrosis b. Mosaic

c. Chlorosis d. Scorching

512. Chemical which prevent the growth and multiplication of fungi is called as

a. Fungicides b. Fungistatic

c. Pesticide d. Bactericides

513. The pathogen of Irish potato famine was

a. *Phytophthora infestans* b. *Rhizoctonia solani*

c. *Alternaria solani* d. *Plasmopara viticola*

514. Somatic phase of the fungi is known as

a. Thallus b. Thallamus

c. Thalli d. Thalakoid

515. Fungi which requires two different host plants to complete its life cycle

a. Autoecious. b. Heteroecious

c. Homothallic d. Microcyclic

516. Yellowing of plants due to lack of light

a. Etiolation b. Chlorosis

c. Mosaic d. Leaf spot

517. The wide spread and severe outbreak of a disease over an area

a. Epidemic b. Pandemic

c. Endemic d. None of these

518. First chemical nematicide came in

a. 1960 b. 1940

c. 1950 d. 1970

519. In case of citrus canker disease, falling of leaves is due to

a. Increase auxin b. Decrease gibberellin

b. Increase ethylene c. Decrease cytokinins

520. First scientist who describes the plant parasitic nematodes

a. P.A. Micheli b. M. Tillet

c. T. J. Needham d. M.J. Barkely

521.According to the recent taxonomic classification citrus canker is caused by

a. *Xanthomonas citri* b. *Xanthomonas axonopodis* pv citri

c. *Xanthomonas compestris citri* d. *Pseudomonas citri*

522. Prion consists of

a. Infectious proteins b. Infectious RNA

c. Both proteins and RNA d. RNA and DNA

523. Genetics of plant pathogens was first studied by
   a. E.C. Stakman b. H.H. Flor
   c. Biffin d. Anton De Bery
524. The bacterium in case of bacterial blight of rice
   a. Stomata b. Enters through Through insect wounds
   c. Hydathodes d. All of the above
525. The ability of the plants to sustain the effect of injury with no or little effecto n yield
   a. Resistant b. Tolerance
   c. Economic threshold level d. Resistance
526. The colony color of *Xanthomonas* is
   a. Yellow b. Light green
   c. Brown d. Orange
527. Bacteria can be best grown on
   a. PDA b. NA
   c. Enrichment media d. Selective media or NA
528. Association of organisms when one injures other for food or other demand
   a. Mutualisms b. Antagonisms
   c. Parasitisms d. Symbiosis
529. Many spores of soil borne fungi fail to germinate due to
   a. Fungicidal activity b. Fungi static
   c. Suppressive soils d. None of these
530. Infection cycle starts with
   a. Pre-penetration b. Penetration
   c. Recognition d. Attachment
531. Virus and Viroids themselves release substances during infection
   a. Enzymes b. Toxins
   c. Growth regulators d. None of these
532. The cuticular waxes are degraded by *Puccinia hordei* by the release of
   a. Enzymes b. Polysaccharides
   c. Mechanical forces d. Toxin
533. Cutin breakdown by cutinase enzymes produces
   a. Monomers b. Oligomers
   c. Both monomers and oligomers d. Sugar
534. The disintegration of starch by the activity of plant pathogens is by the
   a. Lipases b. Phospholipases
   c. Amylases d. Cellulases

535. HV toxin, HC toxin Victorin are the examples of

a. Host specific b. Non host specific
c. Enzymes d. Polysaccahrides

536. In a disease corn smut caused by *Ustilago* maydis, there is increase level of

a. IAA b. Gibberellin
c. Ethylene d. Cytokinins

537. In the diseased plants the rate of photosynthesis

a. Increases b. Decreases
c. Remain normal d. Both increases and decrease

538. The upward and downward movement of water and nutrients is blocked in diseases

a. Root rot diseases b. Bacterial and fungal vascular wilts
c. Leaf spot d. Canker on stem

539. The rate of respiration in the diseased plants

a. Increases b. Decreases
c. Remain stable d. None of these

540. Various substances released by the pathogens in their immediate environment for identification of host

a. Glycoproteins b. Carbohydrates
c. Fatty acid d. All of these

541. Formation of abscission layer is an example of

a. Histological defense structure b. Cell wall defense structure
c. Cytoplasmic defense structure d. Hypersensitive defense structure

542. Hypersensitive (HR) response appear in

a.) Resistant cultivars b. Susceptible cultivars
c.) Both in resistant and susceptible d. Compatible host

543. Phytoalexins are the substances produce as a result of

a. Microorganisms attack b. Chemical and mechanical stimuli
c. Both pathogens and any type of stimuli d. Naturally without any stimuli

544. Substances produce by the engineered plants due to encoded animal's genes

a. Antibodies b. Phytoalexins
c. Plantibodies d. All of theases

545. Resistance which develops in distance, untreated parts of the plants

a. Local acquired resistance b. Systemic acquired resistance
c. Complete resistance d. Partial resistance

546. In Plant pathogenic bacteria, mechanism of genetic variablity
a. Conjugations b. Transformation
c. Transduction d. All of these

547. Condition in which the pathogens loss its virulence in the culture
a. Non virulent b. Attenuation
c. Passage d. Virulence

548. The identical individual produced asexually from variant
a. Race b. Biotype
c. Strain d. Formae Specialis

549. Horizontal resistance is
a. Stable b. Unstable
c. Partial stable d. None of these

550. A virulence (avr) genes was first identified by
a. Anton De Bary b. Robert Hartig
c. H.H Flor d. Robert Kock

551. New strains of many pathogens appear as result of
a. By passing the resistance a. Mutation
c. Both mutation and genetic recombinationd. Genetic Recombination

552. New crop varieties in our country comes in to field as result of
a. Mass selection b. Conventional breeding
c. Genetic engineering d. All these

553. Common scab of potato disease is favored by pH
a. 5.2 to 8.0 b. 8.0 to 10.0
c. 4.2 to5.2 d. None of these

554. Area under disease progress curve indicate
a. The pattern of epidemic b. No. of lesions
c. Amount, No of diseased plants d. All of these

555. EPIVEN is computer simulation program for
a. Gram blight b. Early blight of potato
c. Apple scab d. Powdery mildew

556. Disease pyramid is now consists of
a. Pathogen, host, environment b. Pathogen, host, environment, time, human
c. Pathogen, host, environment, space d. Pathogen, host, environment, human being

557. To keep the pathogen away from the area of the host is known as
a. Exclusion b. Evasion
c. Eradication d. Protection

558. Example of most common mycoparasitic fungi is
a. *Phytophthora* a. *Pythium*
b. *Trichoderma* d. *Rhizocotina*

559. The term cross protection is used commonly for
   a. Viral diseases b. Bacterial diseases
   c. Fungal diseases d. Nematode

560. The toxic substance in the Bordeaux mixture is
   a. Lime b. Copper
   c. Both lime and copper d. None of these

561. The discovery of Bordeaux mixture is made
   a. France b. Germany
   c. UK d. USA

562. Homothallic are the fungi in which the sexual reproduction takes place in
   a. Single thallus b. Multithallus
   c. Compatible thallus d. Heterothallic

563. Sterole inhibiting fungicides inhibit the synthesis of
   a. Cell wall c. Cell membrane
   d. Cytoplasm e. Mitochondria

564. Motile spore is also known as
   a. Aplanospores b. Chlamydospores
   c. Conidia d. Zoospores

565. Time taken by the new synthesis chemical to become pesticides is
   a. 3 years b. 5 years
   c. 8 years d. 10 years

566. Organisms which obtain their food from dead organic matter
   a. Parasites b. Autotrophs
   c. Herbivorovs d. Saprobes

567. Integrated disease control includes
   a. Use of chemicals b. Use of chemical and physical
   c. Use of cultural and biological d. Combining all these

568. When the haploid thallus alternate with the diploid thallus it is called as
   a. Diplobiontic b. Haplobiontic
   c. Haploid d. Trplobiontic

569. Damping off disease is caused by the members of
   a. Peronosporaceae b. Albuginaceae
   c. Pythiceae d. None of these

570. Late blight of potato requires
   a. Low temperature low humidity b. High temperature and high humidity
   c. Low temperature high humidity d. High temperature low humidity

571. The order in which yeasts are **included are**
   a. Saccharomycetales b. Taphrinales
   c. Zoopagales d. Mucorales
572. Late blight of potato can be controlled by
   a. By the application of Hg fungicides b. No fungicides
   c. Biological control d. Biological control and Cu fungicides
573. Members of family Erysiphaceae causes serious group of diseases
   a. Papery mildews b. Powdery mildews
   c. Downy mildew d. Blight
574. Powdery mildews are
   a. Obligate parasites b. Facultative saprophytes
   c. Obligatory sapropytes d. Facultative parasites
575. Alkaloids which are poison for man and animals are produced by
   a. *Venturia inaequalis* b. *Monilinia fruticola*
   c. *Claviceps purpurea* d. *Ascochyta rabiei*
576. Fungi that are grown on animal dung are known
   a. Coprophillous b. Parasites
   c. Saprophytes d. All of these
577. Mushrooms are member of
   a. Basidiomycetes b. Discomycetes
   c. Pyrenomycetes d. Ascomycetes
578. Most powdery mildew diseases are favoured by
   a. Low temp., drier conditions b. High temp., low humidity
   c. High temp., high humidity d. Low temp., low humidity
579. Bacteriology is the study of
   a. Virus b. Bacteria
   c. Nematode d. Amoeba
580. Rice blast disease is favored by
   a. High nitrogen level b. Prolonged leaf wetness
   c. Temp. of about 20 ^{0}C d. All of these
581. Bacteria reproduced asexually by means of
   a. Fragmentation b. Budding
   c. Binary fission d. Multiplication
582. Scab type symptoms are produced due to
   a. Fungi b. Fungi + Bacteria
   c. Virus d. Nematode

583. Rod shaped bacteria are known as

a. Coccus
b. Bacillus
c. Spirillus
d. None of these

584. Hard resistant and physiologically dormant cells of bacteria are called as

a. Endospores
b. Actospores
c. Zoospores
d. Aplanospores

585. Bacteria which have clusters of flagella on each end

a. Lophotrichate
b. Atrichate
c. Amphitricate
d. Peritrichate

586. *Fusarium oxysporium* causes

a. Wilt
b. Wither tip
c. Leaf curl
d. Root rot

587. Which of the following is Phenerogamic parasite

a. Alternaria
b. Bacillus
c. CLCuV
d. Cuscutta

588. Mycoplasma are the bacteria which **lack**

a. Cell membrane
b. Cell wall
c. DNA
d. All of these

589. Mycoplasmas causes human being disease

a. Tuberculosis
b. Pneumonia
c. Abortion
d. Tension

590. *Erwinia carotovora* subsp. *atroseptica* causes

a. Black wart disease
b. Soft rot of apple
b. Soft rot of potato
d. CVC

591. First book on Plant pathology is written by

a. Anton De Bary
b. Prevost
b. Theophrastus
c. Kuhn

592 Brown Rot of stone fruit is caused by

a. *Monilinia fructicola*
b. *Rhizoctonia solani*
c. *Erwinia amylovora*
d. *Taphrina deformans*

593. Stem rust of wheat is

a. Macrocyclic and heteroecious
b. Microcyclic and autoecious
c. Macrocyclic and autoecious
d. Microcyclic and heterecious

594. The example of L- form bacteria is

a. *Agrobacterium tumefaciens*
b. *Xanthomonas malvacearum*
c. *Pseudomonas solanacearum*
d. *Streptomyces scabies*

595. Phytoplasmas can be grown on

a. PDA
b. Nutrient Agar
c. Media added with horse serum
d. None of these

596. Most of the Phanerogamic parasites lack
   a. True roots b. Chlorophyll
   c. True Root and chlorophyll d. None of these
597. Bacterial blight of cotton is
   a. Soil borne b. Seed borne
   c. Air borne d. Insect borne
598. Molecular Plant Pathology started as real branch of Plant Pathology in
   a. 1956 b. 1946
   c. 1966 d. 1936
599. Prion can infect
   a. Plants b. Animals
   c. Human being d. None of these
600. Koch's postulates are used for the confirmation of
   a. Pathogenicity b. Disease cycle
   c. Parasitism d. Resistance
601. *Claviceps purpuria* makes
   a. Bordeaux mixture b. Benlate
   c. Ergot mixture d. Pekecidine
602. Tikka disease of ground nut is reoccurred in
   a. Lahore b. Thar
   c. Chakwal d. Rawalpindi
603. Gram blight is serious disease in
   a. Dry region b. Humid region
   c. Hot region d. Cooler region
604. Red rot of sugar cane usually attacks on
   a. Thick cane b. Thin cane
   c. Medium cane d. Long cane
605. Ear cockle disease is caused by
   a. Bacteria b. Virus
   c. Fungi d. Nematodes
606. Great reduction in the size of leaf due to attack of pathogens
   a. Atrophy b. Necrosis
   c. Hyperplasia d. Hypertrophy
607. *Amanita* is
   a. Poisonous mushroom b. Edible mushroom
   c. Ornamental mushroom d. Domestic mushroom
608. In Uncinula genera the appendages are
   a. Myceloid b. Bulbous
   c. Hook shaped d. Dichotomous

609. The turgor pressure inside the appressorium is
   a. 40 times than the normal car tire
   b. 20 times than the normal car tire
   c. 10 times than the normal car tire
   d. 25 times than the normal car tire

610. The rate of mutation is
   a. One in one million
   b. Ten in one million
   c. One in one billion
   d. Twenty in one billion

611. Mycology is a science that deals with the study of
   a. Bacteria
   b. Fungi
   c. Nematodes
   d. Virus

612. Each filament of the fungus is called
   a. Hypha
   b. Spore
   c. Conidium
   d. Oospore

613. Fungi get their nourishment from host by
   a. Roots
   b. Rhizomorphs
   c. Rhizoids
   d. Haustoria

614 Fungus that can live only on living host is
   a. Facultative parasite
   b. Obligate parasite
   c. Saprophyte
   d. Parasite

615. An Organism living inside the host cell is
   a. Intercellular
   b. Interacellular
   c. Autocellular
   d. Heterocellular

616. Dissimilar gametes are
   a. Isogametes
   b. Anisogametes
   c. Heterogametes
   d. Homogametes

617. Sexual spore formed by the fusion of dissimilar gamets is
   a. Zoospore
   b. Chlamydospore
   c. Oospore
   d. Planospore

618. When entire thallus of the fungus is converted into one or more reproductive structure is
   a. Eucarpic
   b. Holopcarpic
   c. Pseudocarpic
   d. Mesocarpic

619. When Hyphae break up into component cells which behave as spore is
   a. Arthrospore
   b. Chlamydospore
   c. Zoospore
   d. Aplanospore

620. Fusion of two nuclei is
   a. Plasmogamy
   b. Karyogamy
   c. Di-karyogamy
   d. Mesiogamy

621. Those fungi in which every thallus is sexually self fertile is

a.	Homothallic	b.	Heterothallic
c.	Dioecious	d.	Monoecious

622. A free living mass of protoplasm

a.	Pseudo plasmodium is free living	b.	Plasmodium
c.	Antheridium	d.	None of them

623. A thallus which is multinucleate without a septum is

a.	Coenocytic	b.	Monocentric
c.	Polycentric	d.	Sub centric

624. A completely closed Ascocarp is

a.	Apothecium	b.	Cleistothecium
c.	Perithecium	d.	Pseudothecium

625. Number of Ascospores produced in the Ascus are usually

a.	4-in under	b.	8-in under
c.	Indefinite	d.	Not known

626. The Cell wall of fungi is composed of

a.	Cellulose	b.	Pectin
c.	Both cellulose and pectin	d.	None of the above

627. A black, hard mass of hyphae resistant to adverse environmental conditions is

a.	Sclerotium	b.	Prosenchyma
c.	Chlamydospore	d.	Stroma

628. Anthracnose diseases are associated with asexual fruiting body

a.	Pycnidium	b.	Acervulus
c.	Synemma	d.	Sporodochium

629. A sexual reproduction when fusion of entire contents of gametes takes place is

a.	Gametangial contact	b.	Gametangial copulation
c.	Somatogamy	d.	Spermatization

630. A process in which plasmogam, karyogamy, meiosis and mitotic division take place not at specific place and points in the life-cycle is

a.	Parasexuality	b.	Haploidization
c.	Segregation	d.	D- None

631. A localized wound, necrotic lesion, raised and cor ky is

a.	Dry rot	b.	Wet rot
c.	Canker	d.	Scab

632. Rapid death and collapse of seedlings in the seed bed is

a.	Soft rot	b.	Damping off
c.	Anthracnose	d.	None

633. Excessive enlargement of cortical cells

a. Hypertrophy b. Hyperplasia
c. Gall formation d. None

634. Bulbous appendages are formed in the genus

a. Erysiphe b. Pereonospore
c. Bremia d. Phyllactina

635. Vesicle is the structure in which the zoospore are collected before liberation and it is formed in

a. *Phytophthora* sp. b. *Fusarium* sp
c. *Pythium* sp d. *Rhizoctonia*

636. Conidia are septate transversally and vertically in

a. *Alternaria* sp b. *Helminthosporium* sp
c. *Fusarium* sp d. None of the above

637 Condia are hyaline in the spp. of

a. *Ascochyta* sp. b. *Helminthosporium* sp
c. *Rhizoctonia* d. *Diplodia*

638. Pycnidium, asexual fruiting body is formed in

a. *Ascochyta rabiei* b. *Venturia inaequalis*
c. *Fusarium oxysporum* d. *Plasmodiophora*

639. Soil and root inhabiting fungus is

a. *Rhizoctonia solani* b. *Cercosporidium*
c. *Colletotrichum graminicolum* d. None of the above

640. Favourable conditions for the developments of late blight of potato are

a. High tem/High humidity b. High tem/low humidity
c. Low tem/High humidity d. Low tem/low humidity

641. For the development of a plant disease in epiphytotic condition, responsible factors are

a. Host/Pathogen b. Host/environment
c. Host/environment d. None of the above

642. Families Pythiaceae and Albuginaceae are in included in the order

a. Perenosporales b. Blastocladiales
c. Mucorales d. Zoopagales

643. Rust and smut diseases are included in the class

a. Ascomycetes b. Oomycetes
c. Basidiomycetes d. Deuteronycetes

644. Loose smut of wheat is included in the infection type

a. Seedling penetration b. Blossom penetration
c. Local penetration d. None of the above

645. In case of leaf spot of rice the optimum temperature for leaf infection ranges
a. 10-15 ^{0}C b. 20-30 ^{0}C
c. 30-35 ^{0}C d. 35-40 ^{0}C

646. Apple scab disease was 1st recorded in the year
a. 1629 b. 1719
c. 1819 d. 1919

647 The primary infection in apple scab take place by
a. Conidia b. Ascospores
c. Basidiospores d. Zygospore

648. Leaf spots appearing as concentric rings are formed in
a. Late blight of potato b. Leaf spot wheat
c. Early blight of potato d. Downy mildew of gropes

649. In downy mildew disease the fungus appears in the form of
a. Silky and velvety b. Hard crust
c. Corky d. Dark lesions

650. In downy mildew of bajra, symptoms appear predominantly on
a. Leaves b. Inflorescence
c. Stem d. Roots

651. The primary infection in green ear disease of Bajra is caused by
a. Zoospore b. Oospore
c. Zygospore d. None

652. *Cuscutta reflexa* is a parasite of
a. Roots b. Stem
c. Leaves d. Flowers

653. Kiran of cotton is caused due to deficiency of
a. Magnesium b. Calcium
c. Copper d. Nitrogen

654. Khaira disease of rice is caused by the deficiency of
a. Calcium b. Manganese
c. Zinc d. Iron

655. Orobanche is a Phanerogamic parasite and grows on the roots of
a. Tobacco b. Cotton
c. Jute d. Rice

656. Soft rot of fruits and vegetables is caused by the production of
a. Enzymes degrading Cell walls b. Auxins
c. Cytokinins d. Ethylene

657. In foot rot of rice the infected plants become taller, as the fungus produces
a. Fusaric acid b. Picolinic acid
c. Gibberellin d. Pysiculasin

658. White rust of crucifers causes the infection
   a. Local type b. Systemic type
   c. Local and systemic type d. Seedling type

659 Powdery mildew of cucurbits is caused by
   a. *Erysiphe cichoracearum* b. *Erysiphe polygoni*
   c. *Oidium mangiferae* d. None of the above

660. The host method of controlling rust diseases is through
   a. Spray fungicides b. Seed dressing fungicides
   c Cultural practices d. Resistant varieties

661. Nematodes have all the systems except
   a. Circulatory b. Reproductive
   c. Digestive d. Nervous

662. The first plant parasitic nematode was recorded in
   a. 1394 b. 1494
   c. 1743 d. 1694

663. Parasitic and non parasitic nematodes are identified on the basis of
   a. Alimentary canal b. Reproductive system
   c. Stylet d. Tail

Answer Key (Plant Pathology)

1 B	2 A	3 A	4 B	5 B	6 B	7 A	8 C	9 A	10 B
11 A	12 A	13 B	14 B	15 A	16 C	17 B	18 C	19 B	20 A
21 B	22 B	23 A	24 D	25 A	26 B	27 A	28 B	29 C	30 A
31 A	32 D	33 A	34 A	35 B	36 B	37 C	38 A	39 A	40 B
41 A	42 B	43 C	44 B	45 C	46 D	47 B	48 B	49 C	50 C
51 D	52 A	53 C	54 B	55 B	56 A	57 D	58 C	59 C	60 A
61 C	62 B	63 A	64 B	65 A	66 D	67 B	68 B	69 C	70 B
71 A	72 B	73 A	74 A	75 B	76 B	77 B	78 A	79 D	80 A
81 D	82 B	83 A	84 C	85 B	86 D	87 B	88 A	89 B	90 B
91 A	92 A	93 C	94 C	95 B	96 D	97 C	98 D	99 C	100 A
101 C	102 D	103 B	104 B	105 D	106 C	107 D	108 D	109 A	110 B
111 C	112 D	113 C	114 C	115 D	116 D	117 D	118 D	119 C	120 C
121 B	122 A	123 B	124 D	125 C	126 D	127 C	128 B	129 D	130 D
131 B	132 A	133 C	134 C	135 C	136 A	137 C	138 C	139 C	140 B
141 C	142 D	143 C	144 B	145 C	146 A	147 C	148 A	149 D	150 B
151 B	152 C	153 C	154 B	155 D	156 C	157 B	158 C	159 B	160 A
161 A	162 C	163 C	164 C	165 D	166 B	167 C	168 B	169 B	170 D
171 B	172 D	173 D	174 D	175 C	176 A	177 D	178 D	179 A	180 A
181 C	182 B	183 A	184 C	185 A	186 B	187 D	188 C	189 A	190 A
191 A	192 D	193 D	194 D	195 C	196 D	197 A	198 A	199 C	200 A
201 D	202 D	203 C	204 D	205 C	206 B	207 A	208 B	209 D	210 A

211 A	212 C	213 B	214 A	215 A	216 B	217 C	218 C	219 B	220 C
221 B	222 C	223 A	224 D	225 B	226 C	227 A	228 B	229 A	230 C
231 D	232 A	233 A	234 D	235 B	236 A	237 B	238 A	239 A	240 B
241 B	242 A	243 A	244 A	245 A	246 A	247 B	248 B	249 B	250 A
251A	252 B	253 B	254 B	255 A	256 A	257 C	258 C	259 A	260 C
261 A	262 C	263 B	264 C	265 A	266 A	267 C	268 B	269 B	270A
271 A	272 B	273 C	274 C	275 A	276 C	277 B	278 D	279 D	280 D
281 A	282 A	283 A	284 B	285 A	286 A	287 A	288 A	289 C	290 A
291 B	292 A	293 A	294 A	295 A	296 A	297 A	298 D	299 B	300 C
301 B	302 D	303 C	304 A	305 A	306 C	307 A	308 A	309 B	310 D
311 A	312 D	313 B	314 D	315 C	316 A	317 C	318 B	319 A	320 D
321 D	322 D	323 C	324 A	325 D	326 A	327 D	328 B	329 C	330 C
331 A	332 C	333 B	334 D	335 C	336 B	337 A	338	339 D	340 B
341 B	342 D	343 A	344 C	345 D	346 A	347 A	348 C	349 C	350 D
351 A	352 B	353 A	354 C	355 B	356 C	357 A	358 B	359 D	360 B
361 C	362 B	363 D	364 A	365 B	366 D	367 A	368 C	369 B	370 D
371	372 C	373 B	374 B	375 B	376 A	377 D	378 C	379 A	380 D
381 D	382 A	383 D	384 C	385 D	386 A	387 B	388 C	389 B	390 C
391 A	392 D	393 B	394 B	395 B	396 B	397 B	398 C	399 B	400 A
401 A	402 C	403 C	404 B	405 B	406 C	407 D	408 A	409 C	410 B
411 B	412 C	413 D	414 D	415 B	416 C	417 D	418 A	419 D	420 B
421 C	422 C	423 B	424 D	425 B	426 B	427 A	428 A	429 D	430 B
431 C	432 B	433 C	434 B	435 D	436 C	437 A	438 B	439 C	440 A
441 B	442 C	443 B	444 D	445 B	446 B	447 B	448 B	449 B	450 C
451 B	452 B	453 A	454 A	455 B	456 B	457 C	458 B	459 B	460 C
461 D	462 C	463 B	464 C	465 B	466 A	467 A	468 C	469 B	470 B
471 B	472 A	473 B	474 D	475	476 B	477 C	478 A	479 C	480 D
481 C	482 B	483 A	484 D	485 B	486 B	487 B	488 C	489 A	490 B
491 C	492 B	493 D	494 B	495 D	496 D	497 A	498 B	499 A	500 B
501 B	502 D	503 C	504 B	505 C	506 C	507 B	508 A	509 C	510 C
511 C	512 B	513 A	514 A	515 B	516 A	517 A	518 C	519 C	520 C
521 B	522 A	523 B	524 D	525 B	526 A	527 D	528 B	529 C	530 A
531 D	532 A	533 C	534 C	535 A	536 A	537 B	538 B	539 A	540 D
541 A	542 A	543 C	544 C	545 B	546 D	547 B	548 B	549 A	550 C
551 C	552 D	553 A	554 D	555 C	556 B	557 A	558 C	559 A	560 B
561 A	562 A	563 B	564 D	565 C	566 D	567 D	568 A	569 C	570 C
571 A	572 D	573 B	574 A	575 C	576 A	577 A	578 A	579 B	580 D
581 C	582 B	583 B	584 A	585 A	586 A	587 D	588 B	589 A	590 C
591 C	592 A	593 A	594 A	595 D	596 A	597 B	598 A	599 B	600 A
601 C	602 C	603 B	604 A	605 D	606 A	607 A	608 C	609 A	610 A
611 B	612 A	613 D	614 B	615 B	616 C	617 C	618 A	619 A	620 B
621 A	622 B	623 A	624 B	625 B	626 D	627 A	628 B	629 B	630 A
631 C	632 B	633 A	34 D	635 A	636 A	637 D	638 A	639 A	640 C

641 D	642 A	643 C	644 B	645 B	646 C	647 B	648 C	649 A	650 B
651 B	652 B	653 D	654 C	655 A	656 A	657 C	658 C	659 A	660 D
661 A	662 C	663 C							

Quick Revision MCQ

1. Fermentation of sugar occurs by
   a. Mucor b. Saccharomyces
   c. Rhizopus d. Penicillium
2. Mushroom is
   a. Plant consisting of fine grass threads b. An edible fungus
   c. Bryophytes d. Flowering plants
3. Bacteria reproduced asexually by means of
   a. Binary fission b. Budding
   c. Fragmentation d. None of these
4. Koch proposed "Koch's Postulates", which are used to confirm the pathogenicity in the year.
   a. 1880 b. 1884
   c. 1782 d. 1780
5. Conidia are
   a. Sexual spore b. Resting spore C
   c. Asexual spore d. Macroscopic
6. Bengal famine in India was caused due to the disease
   a. Black spot of rice b. Blue spot of leaves
   c. Rust of wheat d. Brown spot of rice
7. Apple scab epidemic in HP was caused due to
   a. *Venturia inaequalis* b. Fusarium oxysporum
   c. Phytophthora infestans d. None of the above
8. Which fungal group is often referred to as the "bread molds"
   a. Chytridiomycota b. Zygomycota
   c. Glomeromycota d. Basidiomycota
9. What is the primary function of fungal cell walls?
   a. To store energy b. To facilitate reproduction
   c. To provide structural support d. To protect against dehydration
10. -------------- are eukaryotic, achlorophyllous, spore bearing organism with absorptive nutrition which generally reproduce both sexually and asexually and whose somatic structure is known as hypha surrounded by a cell wall composed of chitin and glucans
    a. Bacteria b. Lichens
    c. Virus d. Fungi

11. He experimentally proved that the *Phytopthora infestans* is the cause of the Late blight of potato.
    a. Micheli b. Fries
    c. Anton de Bary d. None
12. Most prevalent rust in India is
    a. Cedar apple rust b. Brown rust of wheat
    c. Yellow rust of wheat d. Black rust of wheat
13. It is a localized self-induced cell death at the site of infection of a host plant by pathogen
    a. Hypersensitivity b. Para sexuality
    c. Heterothallism d. None of the above
14. A whitish superficial growth on the upper surface of the leaves and stems is due to
    a. Downy mildew b. Blight
    c. Powdery mildew d. Rust
15. The elements of disease tetrahedron are host, pathogen
    a. Environment and disease b. Environment, time and human
    c. Time and human interference d. None
16. Any agent (living organism or virus) which can cause disease.
    a. Mucor b. Fungus
    c. Fungi d. Pathogen
17. The ability of an organism to exclude or overcome completely or in some degree, the effect of a pathogens, or other damaging factor.
    a. Resistance b. Pathogenesis
    c. Hypersenstivity d. Virulence
18. Plant wax is a
    a. Post induction structure b. None of the above
    c. Resistance d. Pre existing defense structure
19. Formation of cork layer in cells is due to
    a. Pre existing defense structure b. Resistance
    c. Post infection defense d. None
20. Tyloses are-formed due to infection of
    a. Viruses b. Leaf spots
    c. Blight pathogens d. Wilt pathogens
21. Glucanases and pectinases produced by pathogens cause
    a. wilts b. Breakdown of cell wall
    c. Heart Rots d. Butt rot

22. Fungistatic compounds produced by plants in response to injury (mechanical or chemical) or infection.
    a. Proteins — b. Toxins
    c. Phytoalexins — d. Gums
23. Dynamics of change in plant disease in time and space is
    a. Virulence — b. Hypersenstitivity
    c. Disease — d. Epidemic
24. Sporadic diseases occur in
    a. Moderate form occasionally — b. Severe form occasionally
    c. Severe form regularly — d. Moderate form regularly
25. Wheat black stem rust is caused by
    a. *Puccinia hordei* — b. *P.graminis* var. *tritici*
    c. *P.graminis* — d. *P. tritici*

Answer key

1.B	2.B	3.A	4.B	5.C	6.D	7.A	8.B	9.C	10.D
11 C	12.C	13.A	14.C	15.B	16.D	17.A	18.D	19.C	20.D
21.B	22.C	23.D	24.A	25.B					

SECTION -II

1. Organ of excretory system of plant parasitic nematode is
    a. Ampulla — b. None
    c. All of the Above — d. Ganglia
2. In plant parasitic nematodes the nerve ring is encircled with
    a. Intestine — b. Bulb
    c. Isthmus — d. Stylet
3. Longest plant parasitic nematode is
    a. Meloidogyne — b. Longidorus
    c. Trichodorus — d. Heterodera
4. Xiphinema , Trichodorus spp. of nematodes Transmit
    a. Mites — b. Spider
    c. Aphids — d. Virus
5. Tundu disease appear in severe form in presence of which nematode
    a. *Anguina* spp. — b. *Heterodera* spp.
    c. *Meloidogyne* spp. — d. *Globodera* spp.
6. Trade name of cerbofuran is
    a. Furadan — b. Thimet
    c. Mancozeb — d. None

7. NETU virus is transmitted by
   a. Paratrichodorus b. Both A and C
   c. Trichodorus d. None
8. Nematode-fungus association was first reported by
   a. Kuhn b. Berkley
   c. De Bary d. Atkinson
9. The most used methods for exclusion of the pathogens are..?
   a. Mechanical b. Regulatory
   c. Chemical d. Biological
10. SAR stands for…?
   a. System acquired resistance b. Systematic aqua resistance
   c. Systematic acquired resistance d. Systemic acquired resistance
11. The internally seed-borne infection can be destroyed by..?
   a. Hot water treatment b. Systemic fungicides
   c. Solar energy treatment d. All of the above
12. PCNB (quintozone) is..?
   a. Non persistent in soil b. Moderate persistent in soil.
   c. Highly persistent in soil. d. All of the above
13. The fungicidal pastes…?
   a. Used in dressing of pruned surface of the tree
   b. Prepared by using copper sulphate plus lime.
   c. Prepared by using copper carbonate plus red lead
   d. All of the above
14. The systemic fungicides have..?
   a. Protective effect b. Curative effect
   c. Niether of A and B d. Both A and B
15. A pathogen often develops resistant against..?
   a. Non systemic fungicides b. Dusts and pastes
   c. Fumigants d. Systemic fungicides
16. Certification is which method of management of pathogen?
   a. Eradication b. Avoiding contact between pathogen and host
   c. Field sanitation d. Exclusion
17. Which of the following is not a symptom of bacterial diseases in plants?
   a. Mosaic b. Blight
   c. Wilt d. Rot
18. Which of the following is not a common measure of controlling viral infections?
   a. Insect control b. Resistant varities
   c. Dusting d. None

19. Which of the given diseases is related to Sugarcane?
    - a. Red rot
    - b. Chlorotic leaf spot
    - c. Gummosis
    - d. Leaf blast
20. Fire blight disease is related to which of the given fruits?
    - a. Pine apple
    - b. Apple
    - c. Orange
    - d. Banana
21. Phyllody, virescence, bunchy top, witches broom are symptoms of
    - a. Virus and bacteria
    - b. Virus and viroid
    - c. Virus and fungus
    - d. Phytoplasma
22. Fruit cracking in tomatoes is due to deficiency of
    - a. Sulphur
    - b. Iron
    - c. Boron
    - d. Nitrogen
23. Whip tail of cauliflower is due to
    - a. Molybdnum
    - b. Sulphur
    - c. Zinc
    - d. None
24. Example of polycyclic disease is
    - a. Wilt
    - b. Rust
    - c. Root rot
    - d. Canker
25. Most prevalent rust in India is
    - a. Cedar apple rust
    - b. Brown rust of wheat
    - c. Yellow rust of wheat
    - d. Black rust of wheat

Answer Key

1.A	2.C	3.B	4.D	5.A	6.A	7.B	8.D	9.B	10.D
11.D	12.C	13.D	14.B	15.D	16.B	17.A	18.C	19.A	20.B
21.D	22.C	23A	24.B	25.C					